BIOLOGICAL APPROACHES FOR CONTROLLING WEEDS

Edited by **Ramalingam Radhakrishnan**

Biological Approaches for Controlling Weeds

http://dx.doi.org/10.5772/intechopen.71593

Edited by Ramalingam Radhakrishnan

Contributors

Alexander Berestetskiy, Sofia Sokornova, Harsh Raman, Nawar Shamaya, James Edward Pratley, Martin Heide Jorgensen, Mozaniel Santana De Oliveira, Wanessa Almeida Da Costa, Priscila Nascimento Bezerra, Antonio Pedro Da Silva Souza Filho, Raul Nunes De Carvalho Junior, Ozoemena Ani, Ogbonnaya Onu, Gideon Onyekachi Okoro, Michael Uguru, Ramalingam Radhakrishnan

Notice

Statements and opinions expressed in the chapters are these of the individual contributors and not necessarily those of the editors or publisher. No responsibility is accepted for the accuracy of information contained in the published chapters. The publisher assumes no responsibility for any damage or injury to persons or property arising out of the use of any materials, instructions, methods or ideas contained in the book.

First published in London, United Kingdom, 2018 by IntechOpen
IntechOpen is the global imprint of INTECHOPEN LIMITED, registered in England and Wales, registration number: 11086078, The Shard, 25th floor, 32 London Bridge Street
London, SE19SG – United Kingdom
Printed in Croatia

British Library Cataloguing-in-Publication Data
A catalogue record for this book is available from the British Library

Additional hard copies can be obtained from orders@intechopen.com

Biological Approaches for Controlling Weeds, Edited by Ramalingam Radhakrishnan
p. cm.
Print ISBN 978-1-78923-654-5
Online ISBN 978-1-78923-655-2

We are IntechOpen,
the world's leading publisher of
Open Access books
Built by scientists, for scientists

3,700+
Open access books available

115,000+
International authors and editors

119M+
Downloads

Our authors are among the

151
Countries delivered to

Top 1%
most cited scientists

12.2%
Contributors from top 500 universities

Interested in publishing with us?
Contact book.department@intechopen.com

Numbers displayed above are based on latest data collected.
For more information visit www.intechopen.com

Meet the editor

Dr. Ramalingam Radhakrishnan was born in India in 1982. He has received several research awards and fellowships during Doctor of Philosophy and made a significant contribution to the *application of magnetic field on improvement of crop plants*. His research was honored by the Chinese Academy of Science by giving financial support to present his findings in an international conference held in China. Professionally, he was employed as a post-doctoral researcher, research professor and assistant professor in different South Korean universities and research institutes. His major research finding is the utilization of microbes or elicitors to improve crop plant growth under environmental stress conditions and biological weed control. He has published several research and review papers as the main author in reputed journals, books and conferences.

Contents

Preface IX

Chapter 1 **Introductory Chapter: Need of Bioherbicide for Weed Control 1**
Ramalingam Radhakrishnan

Chapter 2 **Overview of Biological Methods of Weed Control 5**
Ozoemena Ani, Ogbonnaya Onu, Gideon Okoro and Michael Uguru

Chapter 3 **The Effect of Tillage on the Weed Control: An Adaptive Approach 17**
Martin Heide Jorgensen

Chapter 4 **Genetic Variation for Weed Competition and Allelopathy in Rapeseed (Brassica napus L.) 27**
Harsh Raman, Nawar Shamaya and James Pratley

Chapter 5 **Potentially Phytotoxic of Chemical Compounds Present in Essential Oil for Invasive Plants Control: A Mini-Review 49**
Mozaniel Santana de Oliveira, Wanessa Almeida da Costa, Priscila Nascimento Bezerra, Antonio Pedro da Silva Souza Filho and Raul Nunes de Carvalho Junior

Chapter 6 **Production and Stabilization of Mycoherbicides 63**
Alexander Berestetskiy and Sofia Sokornova

Preface

Weed populations in agriculture are a major cause of yield loss. Conventionally, crop rotation and tillage practices limit the number of weed flora. Biological agents are used as bioherbicides against weeds, which are an alternative to chemical herbicides. Several studies reveal that plant extracts, bacteria, fungi and their products effectively control weed seed germination and growth.

Biological Approaches for Controlling Weeds is intended to offer current knowledge on biological methods to control weed populations. It includes six chapters. The introductory chapter presents the hazardous effects of chemical herbicides and the need of biological agents to control weed populations. Chapter 2 describes the overall biological approach of weed control, including principles and procedures for biological weed control, success rate of biological agents and making the choice of which agents to use. The conventional way of weed control by using the tillage system is explained in Chapter 3. It shows the importance of tillage types, seed bed preparation and weed removal. Chapter 4 reveals that a genetic difference between weed competition and allelopathy is evidenced with canola populations. Plant extracts and their essential oils suppress weed growth and their phytotoxic activities are focused on in Chapter 5. In Chapter 6, the role of mycoherbicides on weed control, production of fungal spores, propagules, mass cultivation, storage and their utilization are discussed.

The information provided in this book will be useful for researchers, students and farmers to understand the importance of bioherbicides. All the authors are gratefully acknowledged for their efforts in writing the chapters.

Dr. Ramalingam Radhakrishnan
Assistant Professor
Department of Microbiology
Karpagam Academy of Higher Education
Eachanari, Coimbatore, India

Chapter 1

Introductory Chapter: Need of Bioherbicide for Weed Control

Ramalingam Radhakrishnan

Additional information is available at the end of the chapter

http://dx.doi.org/10.5772/intechopen.77958

1. Introduction

Food production is affected by climatic changes and environmental pollutions. The growth and yield of crop plants are significantly declined due to the effect of weed (a plant considered unwanted in a particular location) growth in farming fields. Weeds are strong competitors against crops to the absorption of water and nutrition from the soil, and also occupy more soil area, which result to suppress the crop growth [1, 2]. The integrated approach of weed control management (including tillage, mechanical way of weed removal, and crop rotation) can able to effectively decrease the weed growth [3–5]. The application of chemical-based herbicides, that is, 2,4-dichlorophenoxyacetic acid (2,4-D), glyphosate, and dicamba suppress the germination and growth of weeds, but the prolonged application of those chemicals could not effectively control the weeds and causes to develop the resistant weed germplasms and also pollutes the environment [6]. In addition, Kim et al. [7] reported that 32% of food products in Korea are unsuitable for consumption due to higher accumulation of pesticides. Recently, several biological organisms or their extracts are utilized to integrate weed control strategies [8].

2. Importance of bioherbicides

Bioherbicides are either living organisms or the natural metabolites that have the ability to control weed populations without harming the environment [9, 10]. The numbers of bacterial and fungal species demonstrate their host-specific or nonspecific bioherbicide activities against susceptible weed populations [9]. In 1980, the commercial form of bioherbicide was first introduced in the USA, Canada, Ukraine, and Europe [8, 10, 11]. The microbial agents such as *Alternaria, Bacillus, Chondrostereum, Colletotrichum, Curvularia, Dactylaria, Diaporthe, Drechslera, Enterobacter,*

Epicoccum, Exserohilum, Fusarium, Gloeocercospora, Microsphaeropsis, Mycoleptodiscus, Myrothecium, Phoma, Phomopsis, Plectosporium, Pseudolagarobasidium, Pseudomonas, Puccinia, Pyricularia, Pythium, Sclerotinia, Serratia, Stagonospora, Streptomycetes, Trichoderma, Verticillium, and *Xanthomonas* species and also several plant extracts have been recorded as bioherbicides [12].

Even though numerous plant products and microbes have been successfully showing the positive results against weeds in field trialed, only a few (one plant extract, three bacteria, and nine fungi) of them are commercially available in current markets [8]. Hoagland [13] demonstrated that the crop plants especially tomato produces allelochemicals such as tomatine and tomatidine, which prevent the growth of weeds and pathogenic fungi. Recently, the researchers are interested to know the bioherbicide compounds by extracting DNA fragments obtained from the soil and cloning the genes in vectors to produce phytotoxic compounds [14]. The mode of action of bioherbicide is not well elucidated, but a few studies revealed that the toxic metabolites from the microbes or plant-based products prevent the population of weeds by affecting cell division, pigment synthesis, nutrient uptake, plant growth promoting regulators, antioxidants, stress-mediated hormones, and other metabolites [13]. In organic farming, the bioherbicide approach is used to avoid herbicide resistance and increase crop yield [15]. In this book, the importance of bioherbicides and integrated management of weed control with tillage, mulching, and other eco-friendly methods are enlightened.

Author details

Ramalingam Radhakrishnan

Address all correspondence to: ramradhakrish@gmail.com

Department of Microbiology, Karpagam Academy of Higher Education, Coimbatore, India

References

[1] Arnold RN, Murray MW, Gregory EJ, Smeal D. Effects of herbicides on weeds in field corn grown on coarse-textured soils. Journal of Applied Agricultural Research. 1988;**3**:121-123

[2] Halford C, Hamill AS, Zhang J, Doucet C. Critical period of weed control in no-till soybean and corn (*Zea mays*). Weed Technology. 2001;**15**:737-744

[3] Marshall EJP, Brown VK, Boatman ND, Lutman PJW, Squire GR, Ward LK. The role of weeds in supporting biological diversity within crop fields. Weed Research. 2003;**43**:77-89

[4] Chikowo R, Faloya V, Petit S, Munier-Jolain NM. Integrated weed management systems allow reduced reliance on herbicides and long-term weed control. Agriculture, Ecosystem and Environment. 2009;**132**:237-242

[5] Koocheki A, Nassiri M, Alimoradi L, Ghorbani R. Effect of cropping systems and crop rotations on weeds. Agronomy for Sustainable Development. 2009;**29**:401-408

[6] Green JM, Owen MDK. Herbicide-resistant crops: Utilities and limitations for herbicide-resistant weed management. Journal of Agricultural Food Chemistry. 2011;**59**(11):5819-5829

[7] Kim HK, Choi DS, Kim SG. Analysis of recent four years situation for pesticide residues in the GAP certified agricultural products analyzed by national agricultural cooperative federation. The Korean Journal of Pesticide Science. 2013;**17**(4):271-282

[8] Cordeau S, Triolet M, Wayman S, Steinberg C, Guillemin JP. Bioherbicides: Dead in the water? A review of the existing products for integrated weed management. Crop Protection. 2016;**87**:44-49

[9] Hoagland RE, Boyette CD, Weaver MA, Abbas HK. Bioherbicides: Research and risks. Toxin Reviews. 2007;**26**:313-342

[10] Bailey KL. In: Abrol, Dharam P, editor. The Bioherbicide Approach to Weed Control using Plant Pathogens, Integrated Pest Management: Current Concepts and Ecological Perspective. San Diego: Elsevier (Academic Press); 2014. pp. 245-266

[11] Charudattan R. Biological control of weeds by means of plant pathogens: Significance for integrated weed management in modern agro-ecology. Biological Control. 2001;**46**:229-260

[12] Radhakrishnan R, Alqarawi AA, Abd_Allah EF. Bioherbicides: Current knowledge on weed control mechanism. Ecotoxicology and Environmental Safety. 2018;**158**:131-138

[13] Hoagland RE. Toxicity of tomatine and tomatidine on weeds, crops and phytopathogenetic fungi. Allelopathy Journal. 2009;**23**(2):425-436

[14] Kao-Kniffin J, Carver SM, Di-Tommaso A. Advancing weed management strategies using metagenomic techniques. Weed Science. 2013;**61**:171-184

[15] Cai X, Gu M. Bioherbicides in organic horticulture. Horticulturae. 2016;**2**(2):3

Chapter 2

Overview of Biological Methods of Weed Control

Ozoemena Ani, Ogbonnaya Onu,
Gideon Okoro and Michael Uguru

Additional information is available at the end of the chapter

http://dx.doi.org/10.5772/intechopen.76219

Abstract

Exotic plants in new ecosystems where they may be of no economic importance and where their original biological enemies may be absent become weeds, difficult to manage by crop farmers. They limit the productivity of the lands and hence affect crop development and yield. Efforts towards reducing reliance on herbicides and other methods for environmental, health, economic and sustainability reasons have led to increasing interest in the biological approach to controlling these weeds. This work therefore presents an overview of the biological approach to weed control with focus on the basic concepts, underlying principles, procedures and current practices, cases and causes of failure and successes. Specifically, this chapter has discussed the underlying principles, general procedures, reasons for relatively slow popularity and adoption of biological weed control, examples of successful biological control of weeds with introduced insects and pathogens, when is weed biological control successful?, things to consider when making the choice of agents to be introduced to control weeds and steps to identifying and introducing biological control agents.

Keywords: biological weed control, overview, weed control methods, biocontrol agents

1. Introduction

Sharp increase in international trades and travels over past decades has led to invasive plants becoming a global problem. Plant invasions cause serious threat to the existence of endangered species and the integrity of ecosystems, which cost national economies tens of billions of dollars every year [1–3]. Weeds have been noted by organic horticulture producers as one of the most expensive, time consuming and troublesome activities in production [4]. Weeds are the most significant of the economic and environmental crop loss factors and much of the weedicides applied all over the world are targeted at them. Invasive weeds cause enormous environmental damage [5]. Also according to [6] weeds disrupt the ecology and the functioning of rangeland plant communities and decrease the quality of services and commodities obtainable from this diverse and important natural resource. In the developing countries, weeding accounts for up to 60% of the total pre-harvest labor input and this is usually by use of simple hand tools [7]. Weeds are generally defined as plants growing where they are not wanted. Popular methods of weed control such as mechanical and chemical are known to be: expensive, energy and labor intensive, require repeated applications, and are unsuitable for managing wide spread plant invasions in ecologically fragile conservation areas or low-value habitats, such as range lands and many aquatic systems. Also mechanical methods cause soil disturbance that may eventually lead to erosion; chemical herbicides cause environmental pollution that pose dangers to human health and wildlife, and certain weed species have developed resistance to some chemical herbicides [1, 8]. Biological approach to weed control dates back from 1795 when *Dactylopius ceylonicus* was introduced to control drooping prickly pear (*Opuntia vulgaris* Miller) over a large area of land; and since then biological control of weeds have been mainly through the classical strategy of introducing natural enemies from areas of co-evolution [9–14].

Biological control agents usually target their specific natural enemy weeds. Recently due to certain favorable environmental [15], health, economic and sustainability reasons; foreign and native organisms that attack weeds are being evaluated for use as biological control agents that may be used to complement conventional methods especially where some weeds have developed resistance to chemical control. Wheeler et al. [16] reported that their international team discovered and tested numerous new species of potential biological control agents that could attack different plant tissues such as defoliators, sap-suckers, stem borers, and leaf- and stem-gall formers. Many successful biological weed control programs in many parts of the world have demonstrated the potency of this approach and support the concept that natural enemies can contribute to the reduction of plant growth and reproduction [17, 18]. Wapshere et al. [19] classified biological approach to weed control as follows: the classical or inoculative method which is based on the introduction of host-specific exotic natural enemies adapted to exotic weeds; the inundative or augmentative method which is based on the mass production and release of native natural enemies usually against native weeds; the conservative method which is based on reducing numbers of native parasites, predators and diseases of native phytophages that feed on native plants; and the broad-spectrum method which is based on the artificial manipulation of the natural enemy population so that the level of attack on the weed is restricted to achieve the desired level of control. According to McFadyen [5] classical method is the predominant method in weed biocontrol. He further explained that classical method involves the introduction and release of agents in form of exotic insects, mites or pathogens to give permanent control, while inundative involves the releases of predators, use of bioherbicides and other integrated pest management which usually are not as widely used as the classical method. Also there are three different techniques for applied biocontrol: (i) conservation—protection or maintenance of existing populations of biocontrol agents; (ii) augmentation—regular action to increase populations of biocontrol agents, either by periodic releases or by environmental manipulation; and (iii) classical biocontrol—the importation and release of exotic biocontrol agents, with the expectation that the agents will become established and further releases will not be necessary.

Louda and Masters [6] stated that despite the positive impact of chemical herbicides in agricultural productivity, complete reliance on chemical control has caused severe problems such as high cost per unit area, decreasing effectiveness, negative impact on plant diversity and increased environmental contamination. He therefore pointed out that the use of biological factors that naturally limit weed populations is one promising alternative. Menaria [20] discussed bioherbicides as an eco-friendly approach to weed management. He explained that the use of chemical herbicides leaves some chemical residues in food commodities which directly or indirectly affect human health. According to him this situation led to the search for alternative methods that are environmentally friendly, and biocontrol has been found a suitable alternative. Green [21] reviewed the potential for control using bioherbicides of four important forest weed species in the UK; including bracken, bramble, Japanese knotweed and rhododendron. They concluded that rhododendron is a suitable target weed for control using wood-rotting fungus as a bioherbicide stump treatment; and this is an approach already developed for weedy hardwood species in South Africa, Canada and Netherlands. Clewley et al. [22] analyzed factors associated with control programs (invasive region, native region, plant growth form, target longevity, control agent guild, taxonomy and study duration) in order to identify patterns of control success. They found out that biological control agents significantly reduced plant size (28 ± 4%), plant mass (37 ± 4%), flower and seed production (35 ± 13 and 42 ± 9%, respectively) and target plant density (56 ± 7%).

2. Underlying principles and procedures for biological weed control

2.1. Underlying principles

The underlying principle behind biological approach to weed control is based on some research works that reported that exotic plants become invasive because they have escaped from the insect herbivores and other natural enemies that limit their multiplication and distribution in their native regions [23–25]; however some other factors may contribute to the tendency for particular plant species to become invasive [26–28]. Therefore biological control involves using specific natural enemies that can diminish the development and reproduction of their prey organism and put some limitations to them [29]. McFadyen [5] stated that the predominant approach to classical biological weed control involves the importation, colonization, and establishment of exotic natural enemies (predators, parasites, and pathogens) to diminish and maintain exotic pest populations to densities that are economically insignificant [30, 31].

2.2. General procedures

Some authors have outlined general procedures to be followed when embarking on classical biological weed control programs as follows: (i) evaluate the ecology, economic impact of the weed and potential conflicts of interest; (ii) survey the organisms that are already attacking the weed in the new habitat in order to distinguish accidentally introduced agents and so eliminate such from future evaluation; (iii) carry out literature search and other forms of survey to identify natural enemies attacking the weed in its native region; (iv) screen the possible biological control agents in the foreign country to determine host range and

Steps	Details
1. Initiation	Data on taxonomy, biology, ecology, economics, native and introduced distributions, known natural enemies, etc., are compiled by initiating scientist or group. An extensive literature review is conducted on the proposed target weed and its relatives, plus known natural enemies. Conflicts of interest identified and resolved if possible
2. Target weed approval	Data in (step 1) submitted to appropriate State and Federal groups for comment; additional data may be required
3. Foreign exploration and domestic surveys	If project approved in (step 2), the center of evolution of the genus of the target weed (if known) and other suitable areas, are searched for natural enemies, particularly where these are eco-climatically similar to the area of introduction. At the same time, the weed should be investigated in the country of introduction for attacking enemies, related plants, etc.
4. Weed ecology and agent host specificity	Ecology of the target weed, its close relatives and its natural enemies is studied in the native area, and the most damaging and apparently selective agents are subjected to several years of host-specificity testing
5. Agent approval	A report on each agent is submitted to appropriate State and Federal bodies to obtain importation and release permits
6. Importation and quarantine clearance	Each agent is imported to the country of introduction where it is reared through at least one generation in quarantine to rid it of its parasites and diseases
7. Rearing and release	After a pure culture of the agent is obtained in (step 6), it is normally mass-reared and released in the field in cages or free at field sites
8. Evaluation and monitoring	Agent is monitored at field sites to determine establishment and degree of stress on target weed, or to determine reasons why the agent did not become established or efficacious
9. Redistribution	To aid spontaneous self-dissemination, agent is distributed to other areas in the target weed's distribution, if needed

Table 1. Summary of steps normally followed when introducing a biological control agent in a classical/inoculative biological control weed program.

specificity, and to remove nonspecific agents from further consideration; (v) carry out further tests of promising candidates in quarantine after introduction to ensure host specificity and eliminate predators, parasites, and pathogens that may have been introduced with them; (vi) embark on mass rearing of host-specific agents; (vii) release the host-specific agents; (viii) carry out post-release evaluation to determine establishment and effectiveness of agents; and (ix) redistribute agents to other areas where control is required [5, 32–34]. Wapshere et al. [19] presented a summary of steps normally followed when introducing a biological control agent in a classical biological control weed program as in **Table 1**.

3. Reasons for relatively slow popularity and adoption of biological weed control

Recent research activities and weed control practices around the world have shown that the old idea derived from untested opinions; that biological approach to weed control is usually very slow, unpredictable, expensive and mostly unsuccessful is totally not true. Apart from the high initial costs, biological approach to weed control has been known to be relatively cheaper when compared to other methods; however certain factors have slowed down the rate of adoption. These factors include: long time of establishment-usually 20 years or more to ensure success, inadequate or no records of the extent of pre-biological control weed infestations that should serve as a guide for a new biocontrol program, discouraging story of poorly implemented weed bio-control programs. A lot of success stories however have been documented [35]. Lack of information about previous successfully implemented biological control of weeds often lead to untested theories becoming established dogma and this negatively influence the decisions to or not apply it [36]. For instance Mcfadyen [35] stated that it was believed that biological control of trees is difficult, but many examples of trees controlled by insects have been reported [37, 38]. Also classical biological control has been viewed as unsuitable for weeds of annual crops or other frequently disturbed environments [39, 40], however there are many examples of successful control of crop weeds [41, 42].

Some researchers have reported that there are evidences showing that some agents introduced for exotic weed control have attacked non target, native plants [43, 44]; and this situation has raised concerns among biological control workers and weed scientists as well as the governments [5, 43, 45, 46]. Opposition to biological approach to control of weeds has also contributed to slowing down the rate of adoption and practice; this is because some researchers and weed control scientists believe that it is difficult to estimate the cost or the feasibility of biocontrol [47]. Based on a study carried out in South Africa, it was reported that some of the weed biocontrol projects have provided practical solutions to problems e.g. the development of Stumpout for the treatment of wattle stumps and the use of *C. gloeosporioides* for the control of *H. sericea*. However other projects have been less successful and have resulted in the rejection of potential agents for various reasons and these include *C. albofundus* on *A. mearnsii*, *X. campestris* on *M. aquaticum and G. nitens* on *R. cuneifolius* [48]. Vurro and Evans [49] identified legislative hurdles, technological and commercial constraints as limitations to the adoption of biological weed control in Europe. Olckers [50] stated that limited budgets in many countries have also helped to slow than the rate of adoption and practice of biological approach to weed control.

4. Examples of successful biological control of weeds with introduced insects and pathogens

One thousand one hundred and forty-four individuals (mostly entomologists and plant pathologists) have ever attended the International Symposia on Biological Control of Weeds (ISBCWs); and out of these, 450–550 weed biological control experts have been actively involved in research and development efforts over the last 50 years mainly from USA, Canada, Australia, South Africa and New Zealand [51]. McFadyen [5] reported that biological approach to weed control has a long history and a good success rate of 94. A comprehensive list of agents and their target weeds have been documented by Winston et al. [52]. Culliney [1] presented potential benefits estimated for some proposed or initiated biological control programs targeting invasive weeds. Frequently cited examples of successful approach to biological weed control are the prickly pear cacti (*Opuntia; spp.*) in Australia, eradicated by an imported moth (*Cactoblastis cactorum*) and rangeland in California, Oregon, Washington, and British Columbia controlled by St. John's wort *Hypericum perforatum* (millepertuis perforé) [53].

Mcfadyen [35] presented a list of 41 weds which have successfully been controlled using introduced insects and pathogens and another three weeds also controlled by introduced fungi applied as mycoherbicides. He further stated that many of these successes have been repeated in other countries and continents. Julien [11] presented a list of both successful and failed cases of biological weed control; this included the introduction of 225 organisms against 111 weed species, and 178 insects and 6 mites. Palmer et al. [54] reported that 43 new arthropod or pathogen agents were released in 19 projects; and that effective biological control was achieved in several projects with the outstanding successes being the control of rubber vine, *Cryptostegia grandiflora*, and bridal creeper, *Asparagus asparagoides*.

4.1. When is weed biological control successful?

Information collated on weed impacts before the initiation of a biological control program is necessary to provide baseline data and devise performance criteria with which the program can subsequently be evaluated [55]. For avoidance of confusion on when a biological control could be viewed as successful or not, Hoffmann [56] stated that an implementation of a particular biological control will be termed successful when: complete-when no other control method is required or used, at least in areas where the agent(s) is established; substantial-where other methods are needed but the effort required is reduced (e.g. less herbicide or less frequent application); and negligible-where despite damage inflicted by agents, control of the weed is still dependent on other control measures. Complete control does not imply total eradication of the weed; rather it means that control measures are not required anymore specifically against the target weed, and that crop or pasture yield losses will not be attributed mainly to this weed [26, 41]. Substantial control involves situations where control may be complete in some seasons and/or over part of the weed's range, as well as cases where the control achieved is widespread and economically significant but the weed is still a major problem. It is therefore concluded that successful implementation of biological approach to weed control is the successful control of the weed, and not necessarily the successful establishment of individual agents released against the weed [35]. Successful biological control depends on three factors: the extent to which each individual agent can limit the targeted plant; the ecology of the agent as it affects its ability to populate and spread easily in the new environment; and the ecology of the weed, which determines if the total damage that can be caused by the agent can significantly reduce its population [57]. Because agents always need some surviving predator plants to complete their life cycle, biological control will not usually totally eradicate their target weeds. In essence a successful biological control program reduces the potency and population of the target weed and usually in conjunction with other control methods as part of an overall integrated weed management scheme which is recommended.

5. Things to consider when making the choice of agents to be introduced to control weeds

Gassmann [58] reported that selection of potential agents in the last decades has been mainly based on the population biology of the weed, impact studies of agents on the plant and the combined effect of herbivory and plant competition. Palmer et al. [54] stated that agent selection is highly dependent on the type of weed, its reproductive system, on the ecological, abiotic and management context in which that weed occurs, and on the acceptable goals and impact thresholds required of a biological control program. Generally, factors to be considered in selecting agents include the following: the agent must target a particular plant species, must have high level of predation and parasitism on the host plant and its entire population, must be prolific, must be able to thrive in all habitats and climates where the weed exists and should be able to spread easily and widely, must be a strong colonizer, the overall cost of introducing the agent must be cheaper compared to other control methods, the technology that will be involved in introducing and managing the agent must be as simple as possible, must as much as possible maintain natural biodiversity, sufficient number of individuals must be released, plant phenology (effect of periodic plant life cycle events) must be favorable [59]. To be considered a good candidate for biological control, a weed should be non-native, present in numbers and densities greater than in its native range and numerous enough to cause environmental or economic damage, the weed should also be present over a broad geographic range, have few or no redeeming or beneficial qualities, have taxonomic characteristics sufficiently distinct from those of economically important and native plant. Furthermore, the weed should occur in relatively undisturbed areas to allow for the establishment of biological control agents, cultivation, mowing and other disturbances can have a destructive effect on many arthropod biocontrol agents. Inundative biocontrol agents such as bacteria and fungi are less sensitive to these types of disturbances so may be used in cropland.

5.1. Steps to identifying and introducing biological control agents

The study of insect attributes and fitness traits, the influence of plant resources on insect performance, and the construction of comparative life-tables, are the first steps towards an improvement of the success rate of biological weed control [58]. Generally, steps to identifying and introducing biological control agents include: (i) identify target weeds; (ii) identify control agents and determine the level of specialization; (iii) apply controlled release of the agents; (iv) apply full release and determine optimal release sites; (v) for the case of classical methods, monitor release sites; (vi) apply redistribution for the case of classical methods (vii) and maintain control agent populations.

6. Conclusion

The following conclusions are drawn from this study:

i.   In recent times, biological and integrated weed control is gaining popularity over the traditional methods of mechanical and chemical because the latter have been noted to be more expensive, energy and labor intensive and require repeated applications.

ii.   Mechanical methods cause soil disturbance and possible erosion while chemical herbicides lead to pollution of the environment and the aftermath

iii.   Some weed species have developed resistance to some chemical herbicides and biological control readily comes as a viable alternative

iv.   Classical method of biological weed control has been the most popular and widely adopted and practiced; it involves the introduction and release of agents in form of exotic insects, mites or pathogens to give permanent control

v.   Inundative method of biological weed control involves the releases of predators, use of bioherbicides and other integrated pest management which usually are not as widely used as the classical method.

vi.   Biological weed control is presently widely adopted in the USA, Canada, Australia, South Africa and New Zealand.

vii.   The biological approach to weed control holds great prospects for sustainable, environmentally friendly and economically viable control of exotic weeds and should be explored further through research, development and legislation.

Author details

Ozoemena Ani[1]*, Ogbonnaya Onu[1], Gideon Okoro[1] and Michael Uguru[2]

*Address all correspondence to: ozoemena.ani@unn.edu.ng

1 Department of Agricultural and Bioresources Engineering, University of Nigeria, Nsukka, Enugu State, Nigeria

2 Department of Crop Science, University of Nigeria, Nsukka, Enugu State, Nigeria

References

[1] Culliney TW. Benefits of classical biological control for managing invasive plants. Critical Reviews in Plant Sciences. 2005;**24**:131-150

[2] Gerber E, Schaffner U, Gassmann A, Hinz H, Seier M, Müller-Schärer H. Prospects for biological control of *Ambrosia artemisiifolia* in Europe: Learning from the past. Weed Research. 2011;**51**:559-573

[3] Carruthers RI. Biological control of invasive species, a personal perspective. Conservation Biology. 2004;**18**:54-57

[4] Cai X, Gu M. Bioherbicides in organic horticulture. Horticulturae. 2016;**2**:3

[5] McFadyen REC. Biological control of weeds. Annual Review of Entomology. 1998;**43**:369-393

[6] Louda SM, Masters RA. Biological control of weeds in Great Plains rangelands. Great Plains Research. 1993:215-247

[7] Webb M, Conroy C. The Socio-economics of weed control on smallholder farms in Uganda. Brighton Crop Protection Conference Weeds: Brit Crop Protection Council; 1995. pp. 157-162

[8] Harding DP, Raizada MN. Controlling weeds with fungi, bacteria and viruses: A review. Frontiers in Plant Science. 2015;**6**:659

[9] Phatak SC, Callaway MB, Vavrina CS. Biological control and its integration in weed management systems for purple and yellow nutsedge (Cyperus rotundus and C. esculentus). Weed Technology. 1987;**1**:84-91

[10] Goeden RD. Introduced parasites and predators of arthropod pests and weeds. Part 11: Biological control of weeds. U.S. Dep. Agric. Handb. 1978

[11] Julien MH. Biological Control of Weeds, a World Catalogue of Agents and their Target Weeds. Sough, UK: Common. Agric. Bur. Farnham Royal.; 1982

[12] Rao VP, Ghani MA, Sankaran T, Mathur KC. Review of biological control of insects and other pests in south- East Asia and Pacific region. Commonwealth Agricultural Bureaux; Commonwealth Institute of Biological Control. 1971;**6**:59-95

[13] Tyron H. The wild cochineal insect, with reference to its injurious action on prickly pear (*Opuntia* spp.) in India etc. and to its availability for subjugation of this plant in Queensland and elsewhere. Queensland Agricultural Journal. 1910;**25**:188-197

[14] Muller-Scharer H, Scheepens P. Biological control of weeds in crops: A coordinated European research programme (COST-816). Integrated Pest Management Reviews. 1997;**2**:45-50

[15] Sodaeizadeh H, Hosseini Z. Allelopathy an environmentally friendly method for weed control. In: International Conference on Applied Life Sciences (ICALS2012); September 10–12, 2012; Tokey: Faculty of Natural Resources and Desert Study, Yazd University, Iran Management and Planning Organization, Yazd County, Iran; 2012

[16] Wheeler GS, Kay FM, Vitorino MD, Manrique V, Diaz R, Overholt WA. Biological control of the invasive weed *Schinus terebinthifolia* (Brazilian peppertree): A review of the project with an update on the proposed agents. Southeastern Naturalist. 2016;**15**:15-34

[17] Huffaker CB, Kennett C. A ten-year study of vegetational changes associated with biological control of Klamath weed. Journal of Range Management. 1959;**12**:69-82

[18] DeBach P. Biological control of insect pests and weeds. 1964

[19] Wapshere AJ, Delfosse ES, Cullen JM. Recent developments in biological control of weeds. Crop Protection. 1989;**8**:227-250

[20] Menaria BL. Bioherbicides: An eco-friendly approach to weed management. Current Science. 2007:92

[21] Green S. A review of the potential for the use of bioherbicides to control forest weeds in the UK. Forestry. 2003;**76**:285-298

[22] Clewley GD, Eschen R, Shaw RH, Wright DJ. The effectiveness of classical biological control of invasive plants. Journal of Applied Ecology. 2012;**49**:1287-1295

[23] Keane RM, Crawley MJ. Exotic plant invasions and the enemy release hypothesis. Trends in Ecology & Evolution. 2002;**17**:164-170

[24] McEvoy PB. Insect-plant interactions on a planet of weeds. Entomologia Experimentalis **et** Applicata. 2002;**104**:165-179

[25] Hoddle MS. Restoring balance: Using exotic species to control invasive exotic species. Conservation Biology. 2004;**18**:38-49

[26] McEvoy P, Cox C, Coombs E. Successful biological control of ragwort, Senecio jacobaea, by introduced insects in Oregon. Ecological Applications. 1991;**1**:430-442

[27] Hierro JL, Callaway RM. Allelopathy and exotic plant invasion. Plant and Soil. 2003;**256**:29-39

[28] Zedler JB, Kercher S. Causes and consequences of invasive plants in wetlands: Opportunities, opportunists, and outcomes. Critical Reviews in Plant Sciences. 2004;**23**:431-452

[29] Harper DB. Fungal degradation of aromatic nitriles. Enzymology of CN cleavage by *Fusarium solani*. Biochemical Journal. 1977;**167**:685-692

[30] Goeden RD. Biological control of weeds. Introduced Parasites and Predators of Arthropods Pests and Weeds: A World Review. USDA Agriculture Handbook. 1978. pp. 357-14

[31] Harley KLS, Forno IW. Biological Control of Weeds: A Handbook for Practitioners and Students. Inkata Press; 1992

[32] Andres LA, Davis CJ, Harris P, Wapshere AJ. Biological Control of Weeds 1976. pp. 481-499

[33] Batra SW. Insects and fungi associated with *Carduus thistles* (Compositae): The administration; 1981

[34] Wapshere AJ, Delfosse ES, Cullen JM. Recent Developments in Biological Control of Weeds. Canberra, ACT, Australia: CSIRO Division of Entomology; 1989. p. 8

[35] McFadyen REC. Successes in biological control of weeds. In: Spencer NR, editor. Proceedings of the X International Symposium on Biological Control of Weeds; Montana State University, Bozeman, Montana, USA. 2000. pp. 3-14

[36] Chaboudez P, Sheppard AW. Are particular weeds more amenable to biological control? A reanalysis of mode of reproduction and life history. In: Proceedings of the Eighth International Symposium on Biological Control. 1995. pp. 95-102

[37] Dennill G, Donnelly D. Biological control of *Acacia longifolia* and related weed species (Fabaceae) in South Africa. Agriculture, Ecosystems & Environment. Melbourne and Sidney: Inkata Press; 1991;**37**:115-135

[38] Mack RN. Predicting the identity and fate of plant invaders: Emergent and emerging approaches. Biological Conservation. 1996;**78**:107-121

[39] Duke SO. Weed science directions in the USA: What has been achieved and where the USA is going. Plant Protection Quarterly. 1997;**12**:2-6

[40] Reznik SY. Classical biocontrol of weeds in crop rotation: A story of failure and prospects for success. In: Proceedings of the IX International Symposium on Biological Control of Weeds: University of Capetown Rondebosch, South Africa; 1996. p. 503-506

[41] Chippendale JF. The biological control of Noogoora burr (Xanthium occidentale) in Queensland: An economic perspective. In: Proceedings of the VIII International Symposium on Biological Control of Weeds; Melbourne, Australia: DSIR/CSIRO; 1995. pp. 185-192

[42] Marsden JS, Martin GE, Parham DJ, Risdill-Smith TJ, Johnston BG. Skeleton Weed Control. Returns on Australian Agricultural Research. Canberra: CSIRO Division of Entomology; 1980. pp. 84-93

[43] Pemberton RW. Predictable risk to native plants in weed biological control. Oecologia. 2000;**125**:489-494

[44] Suckling DM. Benefits from biological control of weeds in New Zealand range from negligible to massive: A retrospective analysis. Biological Control. 2013;**66**:27-32

[45] Sheppard AW, Hill R, DeClerck-Floate RA, McClay A, Olckers T, Quimby PC, et al. A global review of riskbenefit-cost analysis for the introduction of classical biological control agents against weeds: A crisis in the making? Biocontrol News Info. 2003;**24**:91-108

[46] Funasaki GY, Lai P-Y, Nakahara LM, Beardsley JW, Ota AK. A review of biological control introductions in Hawaii: 1890 to 1985. In: Proceedings of the Hawaiian Entomological Society. 1988;**28**:105-160

[47] Paynter Q, Fowler SV, Hayes L, Hill RL. Factors affecting the cost of weed biocontrol programs in New Zealand. Biological Control. 2015;**80**:119-127

[48] Morris M, Wood A, Den Breeÿen A. Plant pathogens and biological control of weeds in South Africa: A review of projects and progress during the last decade. African Entomology Memoir. 1999;**1**:129-137

[49] Vurro M, Evans H. Opportunities and constraints for the biological control of weeds in Europe. In: Proceedings of the XII International Symposium Biological Control Weeds; Wallingford: CAB International; 2008. pp. 455-462

[50] Olckers T. Targeting emerging weeds for biological control in South Africa: The benefits of halting the spread of alien plants at an early stage of their invasion: Working for water. South African Journal of Science. 2004;**100**:64-68

[51] Moran V, Hoffmann J. The fourteen international symposia on biological control of weeds, 1969–2014: Delegates, demographics and inferences from the debate on non-target effects. Biological Control. 2015;**87**:23-31

[52] Winston RL, Schwarzländer M, Hinz HL, Day MD, Cock MJW, Julien MH. Biological Control of Weeds: A World Catalogue of Agents and their Target Weeds. 5th ed. Morgantown, West Virginia: Forest Health Technology Enterprise Team; 2014

[53] Nowierski RM. Some basic aspects of biological weed control1. Great Plains Agricultural Council Leafy Spurge. 1984. pp. 23-26

[54] Palmer W, Heard T, Sheppard A. A review of Australian classical biological control of weeds programs and research activities over the past 12 years. Biological Control. 2010;**52**: 271-287

[55] Morin L, Reid AM, Sims-Chilton N, Buckley Y, Dhileepan K, Hastwell GT, et al. Review of approaches to evaluate the effectiveness of weed biological control agents. Biological Control. 2009;**51**:1-15

[56] Hoffmann J. Biological Control of Weeds: The Way Forward, a South African Perspective. 1995

[57] Cullen JM. Predicting Effectiveness: Fact and Fantasy. In: Delfosse ES, Scott RR, editors. Proc. 8th. Int Symp Biol Control Weeds. Melbourne: CSIRO; 1995. pp. 103-109

[58] Gassmann A. Classical biological control of weeds with insects: A case for emphasizing agent demography. In: Proceedings of the IX International Symposium on Biological Control of Weeds: University of Cape Town Rondebosch, South Africa; 1996. pp. 171-175

[59] TIA. Biological control of weeds. Agricultural Research Information Bulletin. Weed Biological Control Pamphlet. Tasmanian Institute of Agriculture TIA; 2008

Chapter 3

The Effect of Tillage on the Weed Control: An Adaptive Approach

Martin Heide Jorgensen

Additional information is available at the end of the chapter

http://dx.doi.org/10.5772/intechopen.76704

Abstract

The tillage systems and performance of the operations have an important impact on the weed control. The primary goal for the tillage is to establish the best possible conditions for the crop establishment and growth under the given conditions as soil texture, moisture and so on. In addition, the tillage system also strongly influences the weed pressure and conditions for weed control. As tillage requires a substantial amount of fuel, and affects the leak of nitrogen and CO_2 from the soil, there is a big motivation in optimizing the tillage operations due to the local conditions in the field. A big challenge is how to sense the local conditions and information that are needed to optimize the tillage system for local treatment and intensity. This chapter focuses on how to optimize the tillage operations in a local adaptive approach aiming at the best possible weed control.

Keywords: tillage, weed control, adaptive, soil fertility

1. Introduction

This chapter describes how the tillage operations contribute to the weed control. It is important to understand that the weed pressure, both perennials and annual germinating species, depends on the common conditions controlled by the cropping system, involving the crop rotation, the soil fertility, nutrient strategy, tillage, and direct control methods as weed harrowing and hoeing. The different actions that contribute to weed control can be considered like filters that favor some plant traits and filter out others. The challenge then is to design the growing system like a system of filters such that all weed species are controlled such that they do not grow unrestrained [1, 2]. By all means, diversity is important in the growing system. As a part of this, it is important that the crop rotation includes different crops, seeded in both spring and autumn. It is important to have in mind that fast-germinating and established crops are highly competitive and contribute substantially to weed control [3]. Another specific element is the benefit of cutting perennial forage grass in the rotation three times or more per year to control quack grass and other root-emerging weeds. With respect to the direct control methods, it is important that the crop rotation allows for space to perform dedicated quack grass control after harvest. The presence of row crops allows one to perform the control by hoeing during the growing season. This potentially gives a good weed control but can also cause substantial problems if the operation fails.

The tillage operations are, in general, divided to be part of the primary tillage or the preparation of the seedbed. The primary tillage is aimed to obtain a good turnover of the plant residuals and to maintain a healthy soil structure. In this multi-oriented context, the demands for the tillage operations can be different if it is related to optimizing the soil fertility or the weed control. An example here is the performance of the moldboard plowing. Aiming for an effective control of perennial weeds demands that the plants are cowered deep in the soil or that they are dried out in a starvation strategy. Whereas the requirements related to the soil fertility can with good conditions be fulfilled by more superficial treatments. In fact, an unnecessary intensive tillage strategy will cause harm to the soil fertility [4, 5]. Therefore, in the operative planning, it is very important to be aware of the actual field conditions for the specific year, and thereby also the infield variations.

The performance of the operations for seedbed preparation and seeding also affect the weed germination and the weed control. The goal of this operation is to do the final leveling of the soil surface and to establish the right structure for the soil aggregates to form the best possible conditions for the seeds to germinate and establish growth. If the time schedule allows, it can be beneficial to perform a weed harrowing prior to the seeding to reduce the density of the first generation of germinating weeds. In the seeding operation, it is off course important to establish the best possible conditions for the seed. A quick and fast germination and establishment of the crop is important to optimize the competitiveness against the weed. In addition, a uniform seeding depth is important for the subsequent weed control, in a way that this enables room for weed harrowing prior to the crop seeds that break through the soil surface [6, 7]. A uniform seeding depth causes a uniform germination and propagation of the crop plants, and thereby the best possible conditions for the following weed control by weed harrowing or hoeing.

All this together makes good sense to involve the principles of precision agriculture, also to support the effectiveness of the contribution to the weed control. This can be site-specific primary tillage, site-specific seedbed preparation, fixed tracking and controlled traffic, implement control in general, and row control of the hoeing process.

2. Primary tillage

The primary tillage aims at maintaining or improving the soil structure and soil fertility. In addition, the strategy and operational execution of tillage greatly affect the weed. The soil structures enable the drainage and water absorption. The porosity for drainage and the water capacity of the upper soil layer is controlled by the microbes and the content of organic matter. This is maintained by the incorporation of fresh organic matter with smooth tillage operations. At the same time, at stable conditions, the planning and performance of the primary tillage must ensure that no increase in the occurrence of the root-emerging weeds occurs. To control root-emerging weeds, the plant must be covered in the soil layer the deeper the better. Normally a depth of approx. 20 cm is recommended. This conflicts with the preferred conditions for the turnover of plant residuals and is supported by the presence of a smooth mix into the soil having access to the oxygen from the air and soil moisture getting into contact with the soil fungal and fauna that catalyze the process. Jacobs et al. [8] test has shown that the best conditions for the turnover of plant residuals are placements in the upper soil layer—0–5 cm. In practical the working depth of approx. 20 cm for the primary tillage is used. The experience is that this gives a good balanced result, just that the operator must be aware, that the working depth must be as shallow as possible. Deeper working depth increases the effect due to weed control but reduces the access to oxygen. Tillage operations may not be overdone in intensity as the operations are highly energy consuming. Also, that the tillage is not only positive. Unnecessary tillage damages the soil structure, this both due to the workability in the seed bed preparation, and the porosity. The challenge for the tillage operations is to support the dynamics of the growing system, not the operation itself.

If it occurs from the monitoring of the fields, that it is necessary to apply a dedicated treatment to reduce the occurrence of root emerging weeds this can be done in more ways. One obvious method is to increase the working depth for the primary tillage operation, and to make sure, that all the residuals are effectively covered deeply under the soil layer. Another more dedicated method is to make space in the crop rotation; this allows for a series of operations in the period after harvest. Here there are in principal two different methods: "drying" and "starving". Convenient conditions allow additionally to cultivate just prior to a period with temperatures below 0°C.

2.1. The type of operation

In Scandinavian countries, as in many other countries, the primary implements used for the primary tillage has been the moldboard plow and the stubble cultivator. Often, the stubble cultivator is used for a shallow operation immediately after harvest to stop the growth and cut the roots of weed plants and to catalyze the contact from the microbiological life to the residues. Hereafter, the strategy is different and highly dependent on the crop rotation and local conditions. If there is a need for a dedicated treatment to reduce the occurrence of perennial weeds, it is generally after harvest that a series of repeated stubble cultivations can be performed [4]. Danish tests show that repeated stubble cultivations in the autumn can reduce the density of perennial weeds up to 90%. Similar results can be seen in a test in Norway and Germany [9, 10]. Under wet conditions in Norway, it has been observed that the best results are obtained by applying the treatment in spring prior to moldboard plowing and seeding. By this method, the time for seeding becomes too late and too costly in yield reduction. In the autumn, as the temperatures are getting lower, the plowing operation is performed. Here, the growth is stopped, and the turnover of the residuals are continued, now integrated into the soil and sowed in some depth.

In areas having problems with erosion and where the farmers want to perform a strategy that is highly focused on soil fertility, deeper cultivation with chisel plows are used instead of the moldboard plowing. Under nonchemical conditions, success rate for this strategy to work is hard to come by. One major challenge is keeping a check on the development of the weed species in a way that it damages the yield and growth conditions for the crops. Although more tests have shown that the most important factor in weed control is the soil fertility and the composition of the crop rotation [11–13], It is shown that when the crops are healthy and the yield is high, it can be acceptable that there is some presence of weed the tests also shows that it is possible to control the weed not to develop uncontrolled. Although the earlier mentioned tests have shown that tillage, especially stubble cultivation or hoeing can be helpful in practical use to control the weed.

2.2. Data and precision agriculture in primary tillage

The system of precision agriculture has been developed during the last three decades. The focus has primarily been on fertilizing and pesticide applications. There has been limited focus on the tillage no matter that there are big potentials both due to savings in cost and energy and in the optimization of the operations. The problem is that the controlling parameters such as "soil fertility" are almost impossible to measure by commercially available sensors or sensor systems. In addition, it is quite difficult to transform to a mathematical model.

Though there are potentially gains both in a planning and graduated intensity over the field that can be performed by use of existing technology. For example, the plowing depth can be controlled by both semi-automated means and automatically [14]. From an overall point of view, the precision-based application can be performed at least due to tree challenges that normally occur locally and delimited on the field:

1. Emerging problems with root-emerging weeds

2. Dense soil with low capacity of water accessible for the crop

3. Compacted soil with reduced efficiency in the drainage

The abovementioned effects can be mapped by manual inspection. In recent years, the use of drones for these types of inspections is developed for commercial use [15, 16]. Since the last approx. 2 decades, the global positioning system (GPS) positioning, the tractor computers, and auto guidance have been commercially developed and are now installed on more than 50% of new tractors. Having the digital application map, it is therefore possible to perform precision-controlled tillage operations. This can be done using selected implements that allow for the wanted adjustments, where the precision-based operations in farm level are introduced in the strategy of utilizing possible benefits and build the necessary profile of knowledge and technology needed to be prepared to utilize the upcoming versions of the new implement prepared for precision applications. One example here is the plow. Here, it is already possible to adjust the plowing depth within a given range using existing technology. New developments [14] show systems that are dedicated for depth control in the individual plowing sections. It is also possible to build on double plow sections in front of the main moldboard that controls if plant residuals primarily are mixed in the lower layer of the plowing profile or if they are placed into the bottom of the profile to optimize the control of root-emerging weeds.

The dedicated strategy in relation to optimizing the soil structure by using existing technology relies more on the operation only for the areas, where the operation is needed. In these situations, it is important to be aware, that the tillage operation is only a part of the solution. The aim here is in a gentle way to loosen the soil as a part of a plan involving more actions that together are aiming to revitalize the soil fertility, that beneficially also involves the introduction of dedicated crops, where the roots actively contribute to soil loosening and applying organic matter from manure, compost, or similar.

The state-of-art research in drone technology is concerned not only with mapping the density of weeds but also determining the species [15, 16]. This information will be beneficial as the growing weed species can be used as indicators for soil fertility, and hereby there are also problems with drainage and water capacity. This is important due to crop growth and yield, but it also plays a very important role in weed management as the fertile soil generates vital and robust grooving crops, that compete effectively with the weed.

Due to this context, it makes sense to pay most attention to the optimization of soil fertility by means of the tillage system and other means.

3. Seed bed preparation and seeding

Ideally, the seed bed preparation must take place some days prior to seeding. This performs the task smoothly in the soil. During the days of rest prior to seeding the soil aggregates stabilize, such that the soil structure after seeding has less risk to slam and potential for erosion. Due to weed control the seed bed preparation prior to the seeding operation has another advantage as it can be used as a false seed bed, initiating weed seeds to germinate, and then removed in the seeding operation. In the planning of crop rotation, it makes good sense to make space for the false seed bed operation prior to the seeding of selected spring crops. More tests have been performed to reduce the pool of weed seeds in the soil. Results show that this is almost impossible [17, 18]. The seeding operation is normally performed by the use of implements, that also involves some tillage in the top layer. For a good establishment of the crop, precise seeding depth is important. This is also an advantage for weed control, as it is possible then to perform a weed harrowing operation just prior to the time where the new seeds breakthrough the soil surface. Hereafter it is important that crops perform a fast and robust establishment in this that the crops benefit from fertile soil, due to access to fertilizers, moisture, and the soil structure. Problems with slammed soil surface restrict the access for oxygen into the soil and thereby inhibit the growth. For more cases, the first establishment of the crop is essential. Due to weed control, it is in this period that the competitive strength of the crop due to weeds is established. In the next section, it is described how the competitive strength for crops improves the possibility of getting good results with weed harrowing. More tests [19, 20] show the importance of the timing in the tillage and weed management operations. Due to competitive characteristics there can be two alternative systems for seeding: one is that the seed is distributed evenly over the area to give the best coverage possible and thereby the best competitive strength against weeds. Danish experiences show good results with this system for crops such as rye and barley. If the crop is less competitive in general, or particularly in the early stage, another system can be favorable, to grow crops in rows this enable the possibility of performing weed harrowing and hoeing. In row cropping, the competitive strength due to weeds in the in-row area has improved substantially, as also the overall resistance for the crop in operations as weed harrowing is improved. The disadvantage of row cropping is that operations for weed control are needed in the inter-row area. Due to yield the crop is not that sensitive. The Danish test has shown that for row distances, up to approx. 18 cm for cereals, there is almost no decrease in yield. Other crops such as rape are even less sensitive to the open row distance.

3.1. Digital tools

Due to seeding there are commercial systems that allow for graduated seed density over the field. The graduation can aim for eliminating locally weak conditions for germination or if an improved coverage is required in the early stage. The control input for this must be given by the operator prior to the operation, as there are no systems available that can sense the input for this. Modern seeding implements are designed to operate with a constant seeding depth, with mechanical means that is adjusted prior to the operation in the field. Ongoing research aims to develop systems that by monitoring the actual working depth are adjusted by a dedicated control system. In the system tested the control systems are operating individually on each seeding section. With such a system in operation I will make sense to expect a coming version designed for adaptive control of the seeding depth, for example, to ensure the access to moister and thereby the best possible conditions for germination.

Systems for automatic change from broad seeding to row cropping are also commercially available. Though not fully flexible, they are designed such that every second seeding unit can be closed. Whit such systems, row cropping can be established only in the areas where it makes sense due to reduced soil fertility or structure. For more reasons, there is a big interest in establishing cover crops. This can be quiet challenging under Nordic conditions as the cover crops in general require early seeding for success. Here, row cropping also enables some good conditions under seeding in the inter-row band prior to harvest. In good conditions cover crops can contribute in stabilizing the growing system due to harvest of nitrogen and controlling weed growth. Though the capacity from cover crops to control weeds is not good enough as it can be used to solve problems, stubble cultivation must be brought into play.

3.2. Weed harrowing and hoeing

To avoid weed problems it is important that all means to control the weed is integrated and optimized together with other elements in the cropping system. The interactions and the connections to the tillage operations and other factors are described in the sections earlier.

The weed harrow is normally designed with a flexible frame mounted with more sections, equipped with a set of long elastic tines that all are in touch with the soil surface. It is commercially available to have systems, where the load on the individual sections is equally distributed by an active control system. The intensity from the tines to the soil surface is controlled by the load on the section and the working angle for the tines. This working angle can also be controlled by a central control system that is normally controlled by the operator. The challenge for the operator is to adjust the settings for the harrow and the operation forward speed such that the damage on the crop is limited and the effect on the weeds is optimized. One important factor here is to make the best use of all factors to optimize the growth difference, such that the crop continuously is bigger and more established than the weed. It is also important that the soil is workable without a slammed and hard surface. In an optimal setup the first treatment is performed approx. 1 week before the seeding process to initiate the germination of the first generation of weed seeds. Hereafter follows a precision seeding as earlier described; this does reset the weeds. Then, just before the crop breaks the soil surface, a weed harrowing operation is again performed. Hereafter a break is needed for the crop to be established such that it sustains a next operation. For this operation the operator needs to pay the most attention to the timing and to the adjustment of the harrow as the best result is achieved by carrying out the operation as early as possible, without damaging the crop and when it is still possible to control the weed. Hereafter two more operations can be performed.

As described the weed harrow is a uniform implement that work in the full working with based on the preconditions that it is possible to establish a difference in the sensitivity for the crop and the weed to the treatment. In comparison to this is the hoe that only operates in the inter-row area without crops. By modern implements the guidance of the hoe can be automatically controlled by a vision-based system that enables the hoe to operate quite precisely in relation to the row. As the competitiveness from the row of crops is bigger and closer to the row, it is important to perform hoeing as close as possible without absolutely damaging the crops. A very precise and dedicated operation can be performed by a hoe equipped with elements such as brushes or other tools that work close to the row. The advantage of the hoe is that it has very high efficiency in the inter-row area.

A normal cropping system does normally include weed harrowing for weed control. The overall effect and weed control can be improved by also introducing the hoe for operations where it can be possible operated. Compared to weed harrowing the hoe is less sensitive on hard surfaces and so on, which means that the effect of hoeing is more reliable. Hoeing can also be an important part of the strategy to bring the system back to normal for areas where the weed has had the chance to develop to a level that is problematic for balance in the system. This can be done by opening the rows to make space for the hoe operations in the problematic areas in the field. The hoe is very efficient in the inter-row area. For the operation close to the row some additional systems can be built [21] such as flexible tines or rotating fingers.

4. Discussion and conclusion

Many farmers and research activities have shown that it is possible to control the weed in cropping systems only by use of mechanical and agronomic means. It is essential that the cropping system is carefully planned and adaptively optimized in relation to the local conditions and challenges. The fertile soil is key both to the growth and to the yield of the crops but also in relation to create the best conditions for successful tillage operations and weed control. One of the core elements is if the cropping system makes space for cutting grass. A perennial grass for feed that is cut three times or more in the growing season contributes substantially to weed control and to the optimizing the soil fertility and hereby the soil structure.

Going a little into the details and interdependency of the different elements and operations in the cropping systems, it appears that there are many balances to be aware of and many optimizations to be made. Here, many new technologies can assist in positive results. In this matter it is important also to follow technical development. Some of the concepts that automatic solutions. This must create value in the operation and the cropping system, but it also prepares the farmer to take in the new technical solutions when they are ready and when they potentially can create value in the individual cropping system.

Author details

Martin Heide Jorgensen

Address all correspondence to: mahej@mmmi.sdu.sk

University of Southern Denmark, Odense, Denmark

References

[1] Melander B, Holst B, Rasmussen IA, Hansen PK. Direct control of perennial weeds between crops e implications for organic farming. Crop Protection. 2012;**40**:36-42

[2] Cardina J, Herms C, Doohan D. Crop rotation and tillage system effects on weed seedbanks. Weed Science. 2002;**50**:448-460

[3] Sardana V, Mahajan G, Jabran K, Chauhan BS. Role of competition in managing weeds: An introduction to the special issue. Crop Protection. 2017;**95**:1-7

[4] Melander B, Munier-Jolain N, Charles R, Wirth J, Schwarz J, Weide R, Bonin L, Jensen PK, Kudsk P. European perspectives on the adoption of nonchemical weed management in reduced-tillage systems for arable crops. Weed Technology. 2013;**27**:231-240

[5] Yagioka A, Komatsuzaki M, Kaneko N, Ueno H. Effect of no-tillage with weed cover mulching versus conventional tillage on global warming potential and nitrate leaching. Agriculture, Ecosystems and Environment. 2015;**200**:42-53

[6] Scherner A, Melander B, Jensen PK, Kudsk P, Avila LA. Reducing tillage intensity affects the cumulative emergence dynamics of annual grass weeds in winter cereals. European Weed Research Society. 2017;**57**:314-322

[7] Nielsen SK, Munkholm LJ, Lamande M, Norremark M, Edwards TC, Green O. Seed drill depth control system for precision seeding. Computers and electronics in Agriculture. 2018;**144**:174-180

[8] Jacobs A, Helfrich M, Dyckmans J, Rauber R, Ludwig B. Effects of residue location on soil organic matter turnover: Results from an incubation experiment with 15N-maize. Journal of Plant Nutrition and Soil Science. 2011;**174**:634-643

[9] Perkrun C, Claupein W. The implication of stubble tillage for weed population dynamics in organic farming. Weed Research. 2006;**46**:414-423

[10] Brandsæter LO, Mangerud K, Helgheim M, Berge TW. Control of perennial weeds in spring cereals through stubble cultivation and mouldboard ploughing during autumn or spring. Crop Protection. 2017;**98**:16-23

[11] Shahzad M, Farooq M, Jabran K, Hussain M. Impact of different crop rotations and tillage systems on weed infestation and productivity of bread wheat. Crop Protection. 2016;**89**: 161-169

[12] Gruber S, Claupein W. Effect of tillage intensity on weed infestation in organic farming. Soil & Tillage Research. 2009;**105**:104-111

[13] Benaragama D, Shirtliffe SJ, Gossen BD, Brandt SA, Lemke R, Johnson EN, Zentner RP, Olfert O, Leeson J, Moulin A, Stevenson C. Long-term weed dynamics and crop yields under diverse crop rotations in organic and conventional cropping systems in the Canadian prairies. Field Crops Research. 2016;**196**:357-367

[14] Nielsen SK, Munkholm LJ, Aarestrup MH, Kristensen MH, Green O. Plough section control for optimised uniformity in primarry tillage. Advances in animal bioscience. Precision Agriculture. 2017;**8**(2):444-449

[15] Huang J, Poe R. Weed Spotting By Drone. Vol. 65. Washington: Agricultural Research; 2017. pp. 1-3

[16] Lambert J P T, Hicks H L, Childs C Z, Frexkleton R P. Evaluating the potential of unmanned aerial systems for mapping weeds at field scales: A case study with *Alopecurus myosuroides*. Weed Research. 2018;**58**:35-45

[17] Murphy S D, Clements D R, Belaoussoff S, Kevan P G, Swanton C J. Promotion of weed species diversity and reduction of weed seedbanks with conservation tillage and crop rotation. Weed Science. 2006;**54**:69-77

[18] Legere A, Samson N. Tillage and weed management effects on weeds in barley-red clover cropping systems. Weed Science. 2004;**52**:881-885

[19] Lyon DJ, Young FL. Integration of weed management and tillage practices in spring barley production. Weed Technology. 2015;**29**:367-373

[20] Cordeau S, Smith R G, Gallandt E R, Brown B, Salon P, DiTommaso A, Ryan M R. Timing of tillage as a driver of weed communities. Weed Science. 2017;**65**:504-514

[21] Pannacci E, Lattanzi B, Tei F. Non-chemical weed management strategies in minor crops: A review. Crop Protection. 2017;**96**:44-58

Genetic Variation for Weed Competition and Allelopathy in Rapeseed (*Brassica napus L.*)

Harsh Raman, Nawar Shamaya and James Pratley

Additional information is available at the end of the chapter

http://dx.doi.org/10.5772/intechopen.79599

Abstract

Rapeseed (canola, *Brassica napus* L.) is the second major oilseed crop of the world and provides a source of healthy oil for human consumption, meal for stock markets and several other by-products. Several weed species afflict the sustainable production and quality of canola. Various agronomic practices such as crop rotation, stubble management (e.g. burning), minimum tillage, application of herbicides and cultivation of herbicide resistant varieties have been deployed to minimise yield losses. There is no doubt that herbicide-tolerant cultivars enable management of weeds which are difficult to control otherwise. However, widespread usage increases the risk of herbicide resistance. This is becoming a major impediment in sustaining high crop productivity. Allelopathic and weed competitive varieties are potential tools to reduce the dependence on herbicides and could be grown to suppress weed growth in commercial canola. Genetic variation and 'proxy' traits involved in both crop competition as well as allelopathy have been reported. Further research is required to link genetic variation in weed competition and allelopathy, and genetic/genomic marker technologies to unravel effective alleles to expand breeding activity for weed interference in canola.

Keywords: canola, allelopathy, weed competition, genetic variation, QTL mapping, genome wide association analysis

1. Introduction

Rapeseed (canola, *Brassica napus* L, 2n = 4X = 38) belongs to the family *Brassicaceae*, which is widely distributed across subtropical to temperate regions. It is thought to be originated as a result of natural hybridisation event between *Brassica rapa* (2n = 2X = 20, genome AA) and

IntechOpen

Brassica oleracea (2n = 2X =18, genome CC) [1]. Rapeseed is a close relative of *Arabidopsis thaliana*, a weed species widely distributed in the Northern hemisphere that diverged from *Brassica* ~20 million year ago [2]. Although rapeseed was domesticated approximately 400 years ago, it has become, in recent decades, the leading oilseed crop worldwide [3], providing about 13% of the world's edible oil supply [4]. In Australia, canola was commercially grown for the first time in 1969 [5]. During the last four decades, the rapeseed industry has expanded exponentially with the development and cultivation of canola quality varieties having less than 2% erucic acid and less than 40 micromoles/g meal glucosinolates as well as resistance to blackleg disease, caused by the fungus, *Leptosphaeria maculans*. Higher grain prices and deployment of high yielding and herbicide tolerant hybrid varieties have further played major roles in its expansion. Currently, canola is the third largest broad-acre crop in Australia and is grown on more than 2.3 million ha [6] in a range of environments (*i.e.* <200 mm to >800 mm rainfall) [5]. Canola is usually sown in rotation with cereal crops such as wheat and barley to manage weeds and diseases of both crop types. Research has shown that canola can increase yields of wheat by up to15% [7].

Several weed species such as wild radish (*Raphanus raphanistrum*), shepherd's purse (*Capsella bursa-pastoris*), capeweed (*Arctotheca calendula*), Indian hedge mustard (*Sisymbrium orientale*), annual ryegrass (*Lolium rigidum*) and Paterson's curse (*Echium plantagineum*) afflict the production of canola. Weeds compete with the canola crop for water and nutrient uptake, and for solar radiation. This results in a reduction in the grain yield as well as in grain quality. Up to 90% reduction in grain yield of canola has been reported under high infestation of wild radish [8]. Improved agronomic practices such as stubble burning, minimal tillage, crop rotation, and application of herbicides provide valuable tools in managing weed populations. The option of manual weeding is not cost-effective for broad-acre crops such as canola. Various herbicide groups (A, B, C, D, I, K, M, and N) are currently used to control weeds in canola [9] . In addition, crop rotations provide the opportunity to rotate herbicide groups and delay the evolution of herbicide-resistant weed populations.

2. Development of herbicide resistant varieties

Several herbicide-tolerant canola varieties marketed as Clearfield™ (CL), Roundup Ready™ (RR), and Triazine Tolerant™ (TT) are currently cultivated to widen the herbicide spectrum for control of weeds in canola and other crops. This strategy has played a major role in transforming the canola industry in Australia. The first TT variety of canola, 'Siren', was developed in 1993. Since then, there has been a continuous supply of open-pollinated as well as hybrid TT varieties for commercial cultivation. Although TT varieties had a 10–15% yield penalty [10] and lower oil content, these varieties have been popular among growers particularly where wild radish has been a problem, accounting for 70% of the cropped area in some states of Australia. These varieties have enabled an effective and cost effective management of common weeds, particularly wild radish, and those which are resistant to Group A and B herbicides. The other herbicide tolerant varieties, RR and CL, do not impose yield penalties.

Canola seems to be particularly vulnerable to competition from broad-leaf weeds as there are limited commercial herbicide options available. The canola industry is thus becoming more and more reliant on the herbicide tolerant varieties to provide control options for these major weeds. Analysis of weed resistance status indicates that key canola weeds in Australia are well known for their multiple herbicide sites of action resistances (**Figure 1**) and so existing herbicide options are either compromised or are likely to be. In recent decades, the heavy reliance on herbicides has led to herbicide resistance in numerous weed species such as annual ryegrass and wild radish with major concern being the increased incidence in particular, to Group M herbicide, glyphosate (Roundup®). Many farmers use glyphosate as a pre-planting herbicide to provide a weed-free seedbed. The advent of Roundup Ready (RR) crop varieties has transformed the use of glyphosate into an in-crop broad spectrum, selective herbicide. As a result, it has become the last herbicide used in the season and so any escapes from that use help to build glyphosate-resistant weed populations in subsequent seasons [11].

Evaluation of the herbicides with the highest number of species for which herbicide resistance has been recorded (**Figure 2**) shows that of the 15 herbicides listed, eight are likely to be utilised in canola production, including Imazamox and Imazethapyr for CL canola, glyphosate for RR canola and atrazine and simazine for TT lines. With the development and commercial cultivation of genetically modified (GM) canola, there is now more flexibility to control a broad-spectrum of weeds through stacking of herbicide tolerant traits. For example, farmers now have access to hybrid varieties which have tolerance to glyphosate and triazines, providing pre-emergence as well as in-crop selective herbicide capability. Unfortunately, this gene stacking strategy for herbicide tolerance has further increased herbicide dependency [13] and is likely to lead to quicker herbicide resistance which in turn unfortunately will reduce weed control options.

Figure 1. Weed species resistance to multiple sites of actions [12].

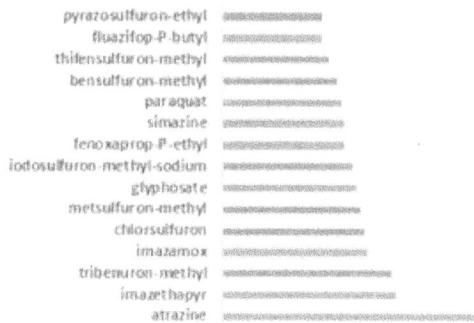

Figure 2. Number of most common resistant species to individual active herbicides (adapted from Heap [12]). Herbicides for use on canola are indicated in orange.

Application of herbicides has its own limitations; the practice is expensive, there is a risk of spray drift to neighbouring crops, and weed resistance threatens the on-going efficacy of the herbicide armoury. An alternative approach is to breed new varieties with improved genetics for weed interference. This interference, which is environmentally friendly, can be of two types: high competitiveness and/or allelopathy. In either case the crop does most of the weed management and herbicides are used in a supplementary way, if at all.

3. Alternative approaches used for weed management: Interference

Crop interference as a tactic has been explored in some crops [14, 15]. It can be defined as the crop plants interfering with weed growth through competition for environmental resources [16] or the crop modifying the growth environment chemically to the disadvantage of the weed [16, 17]. These mechanisms are distinct but seem to act collectively to control weed populations under field conditions [18]. Although allelopathy includes growth promoting, and inhibiting effects, it is usually used to describe growth inhibiting effects [19]. Management practices also can and should assist these processes: for example, growers can manipulate crop sowing times and sowing rates to disadvantage the weeds relative to the crop as well as impose practices that minimise weed seed additions to the seed bank.

3.1. Genetic variation for weed competition

Crop competition is the ability of crops to adapt to weed infestation by accessing limited resources also sought by neighbouring weeds. Traits associated with weed competition are generally related to morphology and phenology of both weed and crop species [20]. Several traits related to competitive ability include plant height, tiller number, leaf angle, canopy structure, early vigour and time to maturity [20]. A good understanding of component attributes underlying those traits would provide an opportunity to improve weed competition of crops using genetic and genomic tools.

Morphological traits related to the interception of radiation by leaves which determine competitiveness for light, including leaf size, number and leaf area index, stem elongation, upward leaf movement [21–24] and leaf layer density [25], have not been studied in canola. These traits are associated with shade avoidance, enabling plants to photosynthesise and grow to improve their competitiveness [21, 22]. Height at maturity has also been reported to contribute to competitive ability [26, 27] although a negative relationship between plant height and weed infestation has been reported for canola [28] and wheat [27]. No such relationship has been found in rice [18]. This trait however tends to have a negative effect on grain yield due to a reduced harvest index.

In wheat, Coleman et al. [29] and Mokhtari et al. [30] showed the normal distribution for phenotypic variation for competitive ability traits in populations derived from crosses between competitive and non-competitive parents. This suggests that the competitive ability trait is controlled by quantitative genes which have minor and moderate effects. Competitive ability associated traits seem to have moderate to highly heritability. In bread wheat, [29] estimated narrow-sense heritabilities for different agronomic and morphological traits associated with weed competition to be: high for flowering date (0.99) and height stem elongation (0.91); low for tiller number (0.34), leaf area index during stem extension (0.18–0.31) and crop dry matter (0.18). Mokhtari et al. [30] estimated the narrow-sense heritability of percentage yield loss due to the weed competition in F_2:F_3 populations of wheat: 0.25 for the population derived from crossing two late flowering time parents and 0.57 for the population derived from crossing between two early flowering time parents.

In rice, broad-sense heritability of weed biomass and crop grain yield under weedy conditions was reported [31] to be high (0.64 to 0.79) for 40 upland rice cultivars grown under weed and weed-free conditions. Another study by Zhao et al. [32] also found that broad-sense heritability was high, being 0.88 for early vigour and 0.81 for crop height 4 weeks after seeding. Although heritability is an indication of phenotypic variation due to genetic effects, the estimation of broad and narrow-sense heritabilities for traits are influenced by population structure and environmental factors.

The genetic bases and extent of variation associated with competitive ability in *Brassica* crops have received attention. In canola, plant height, leaf size, leaf number and leaf area index, stem elongation, upward leaf movement and leaf density are considered as the most important attributes for above ground competition for light; and plant root size and depth, relative growth rate, biomass, root density and total root surface area are the most important traits for below ground competition for space, soil nutrients and water [33]. However, only limited component traits have been studied so far to determine the extent of genetic variation in *Brassica* species. For example, Beckie et al. [34] compared the competitive ability of canola with yellow mustard (*B. juncea*) against wild oats. Yellow mustard was superior in competitiveness to canola due to its rapid growth and plant height resulting in early-season crop biomass accumulation. It has also been shown that canola hybrid varieties are more competitive than open pollinated varieties due to their faster growth and biomass accumulation [35]. Harker et al. [36] confirmed the stronger competitive ability of hybrid canola varieties especially under cool and low growing degree day conditions. In an Australian study, Asaduzzaman et al., (unpublished) compared the weed competiveness of 16 *Brassica napus* genotypes representing open pollinated, F_1 hybrid and TT lines against annual ryegrass and associated weeds and showed that open pollinated and hybrid genotypes reduced weed shoot biomass by 50% compared with less vigorous TT geno-types. In a recent study, Shamaya et al. [37] evaluated the competitive ability of 26 canola genotypes against annual ryegrass (*Lolium rigidum*) under field and glasshouse conditions to study the phenotypic traits associated with weed competition. Under both conditions, the canola biomass, mostly leaf biomass measured in the glasshouse only, was positively associated with competitive ability.

3.2. Detection of QTL for weed competitiveness

Several studies have employed the Quantitative Trait Locus (QTL) mapping approach for detecting, localising and determining the magnitude of loci affecting phenotypic variation for weed competition in plants (**Table 1**). The QTL mapping approach is based on the statistical association between phenotypic and molecular marker polymorphism data. Several molecular markers such as Restriction Fragment Length Polymorphism (RFLP), Single Feature Polymor-phism (SFP), Diversity Arrays Technology (DArTs), Random Amplified Polymorphic DNAs (RAPDS), Simple Sequence Repeats/Microsatellites (SSRs), Amplified Fragment Length Poly-morphisms (AFLPs), Cleaved Amplified Polymorphic Sequence (CAPs) and Sequence-Related Amplified Polymorphism (SRAP) have been used extensively to genotype populations for genetic analyses [38–44]. More recently, whole genome sequencing methods enabled to develop new marker systems such as genotyping by sequencing based on the complexity reduction methods including DArTseq, Single Nucleotide Polymorphisms (SNPs), restriction-site associated DNA (RAD), RNA-Seq and sequence captures that are more suitable for high-throughput analyses [45–50].

Two strategies based on Quantitative Trait Locus (QTL) mapping and genome-wide associa-tion mapping (genome-wide association study, GWAS) approaches have been used to under-stand the genetic basis of natural variation for weed interference in various crop plants such as rice, corn, wheat, cowpea, barrel clover, peas, sorghum, sunflower and *A. thaliana* [51–57]. In *B. napus*, QTL for various traits of agronomic importance including seed germination/plant emergence, fractional ground cover (early vigour), plant biomass, flowering time, plant height, plant maturity, grain yield, resistance to various biotic and abiotic stresses and seed shattering have been mapped using traditional and GWAS [49, 58–74]. However, no QTL associated with weed competition and/or allelopathy has been identified to date.

QTL for weed competition traits have been mapped in cereals and other crops. For example, in wheat Coleman et al. [29] utilised the genetic linkage map based on RFLP, AFLP, SSR, known genes and protein markers of doubled haploid (DH) populations derived from Cranbrook/Halberd to investigate the genetic control of various traits involved in grain yield loss and suppression of ryegrass growth. These traits included the width of the second leaf, canopy height, light interception at early stem elongation, tiller number, days to anthesis and plant height. Several consistent QTL for flag leaf area, flag leaf length, flag leaf width, height at stem elongation, and tiller number were identified in the vicinity of photo-period genes (*Ppd-B1* and

Competitive ability

Species	Population type	Population size	Trait	Season	Chromosome	R^2	Reference
Wheat (*Triticum aestivum* L.)	Doubled haploid lines derived from Cranbrook/ Halberd	161	Yield	1999	3A	12.2	Coleman et al. [29]
					3B	9.8	
			1000 – grain weight	1998	5A	11.0	
					2D	8.4	
				1999	5A	12.0	
					2B	9.9	
Wheat (*Triticum aestivum* L.)	Recombinant inbred lines derived Opata 85/ and synthetic W7984	108	Early Season Vigour	2005	5A	16	Reid [75]
				2006	5A	22	
			Days to Heading	2005	5A	21	
				2006	5A	21	
			Day to Anthesis	2005	5A	20	
				2006	5A	17	
			Days to Maturity	2005	5A	13	
				2006	5A	19	
			Weed Suppression	2005	5A	14	
				2006	5A	15	

Allelopathy

Species	Population type	Population size	Trait	Season	Chromosome	R^2	Reference
Wheat (*Triticum aestivum* L.)	Doubled haploid lines derived from Tasman (strongly allelopathy) Sunco (weakly allelopathy)	271	Reduction in annual ryegrass using the Equal-Compartment-Agar-Method [89]		2B	29	Wu et al., [57]
Rice (*Oryza sativa* L.)	$F_2 - F_3$ population derived from *Indica* line PI312777 (strongly allelopathy) *Japonica* cv Rexmont (weakly allelopathy)	192	Reduction in lettuce root length using water-soluble extract method [116]		1, 3, 5, 6, 7, 11, 12	9.4– 16.1	Ebana et al., [112]
Rice (*Oryza sativa* L.)	Recombinant inbred lines derived from crossing cv IAC 165 (strongly allelopathy) and cv CO39 (weakly allelopathy)	142	Reduction in barnyard grass root length using relay seeding technique method [117]		3	12	Jensen et al., [113]
					3	7.2	
					8	8.5	
Rice (*Oryza sativa* L.)	Doubled haploid lines derived from *Japonica* Jingxi17 (strongly allelopathy) *Indica* Zhaiyeqing 8 (weakly allelopathy)	123	Reduction in lettuce root length using water-soluble extract		3	10.24	Dali et al., [118]
					9	8.24	
					10	8.27	
					12	9.79	
Rice (*Oryza sativa* L.)	Recombinant inbred lines derived from *Indica* cv AC1423 (strongly	150	Reduction in *Echinochloa crus-galli* root length using		4	11.1	Jensen et al., [114]

Competitive ability

Species	Population type	Population size	Trait	Season	Chromosome	R^2	Reference
	allelopathy)/*cv.* Aus196 (weakly allelopathy)		relay seeding technique method [117]				
			Echinochloa crus-galli root length from greenhouse pot set-up		4	9.6	
			Echinochloa crus-galli root biomass from greenhouse pot set-up		3	5.0	
					6	6.9	
			Echinochloa crus-galli shoot length from greenhouse pot set-up		3	5.9	
					8	7.1	
			Echinochloa crus-galli shoot biomass from greenhouse pot set-up		8	5.1	
					12	5.8	
Rice (*Oryza sativa* L.)	Recombinant inbred lines derived from cv. Zhong-156 (strongly allelopathy)/cv. Gumei-2 (weakly allelopathy)	147	Allelopathy index determined by secondary metabolite		11	16.5	Zhou et al. [111]

Table 1. Genetic analysis of mapping populations for crop competitiveness and allelopathy.

Ppd-D1) on the group 2 chromosomes. Three QTL for plant height at anthesis were detected on chromosomes 3A, 4B and 5A. No QTL for crop yield loss in the presence of ryegrass or ryegrass dry matter suppression was identified in this population, likely due to the complex nature of this trait [29]. However this study reported that ryegrass dry matter was suppressed for DH lines of wheat with greater leaf area index, more tillers, taller plant height and later flowering. High genetic correlations between leaf area index and grain yield loss (r = −0.81) as well as suppression of ryegrass (r = −0.91) were observed indicating that traits contributing to early ground cover would be important for developing competitive wheat genotypes. Another wheat study conducted in the northern region of Canada determined a cluster of QTL associated with traits implicated in weed competition [75] using 108 recombinant inbred lines derived from a cross between Mexican wheat, Opata 85, and a synthetic wheat accession, W7984. Early vigour, day to heading, day to anthesis, day to maturity and weed suppression were mapped to the same region on chromosome 5A corresponding to the position of the vernalisation gene *Vrn-A1*, suggesting that flowering time may be associated with weed suppression.

In rice (*Oryza sativa* L), a mapping population developed from a cross between a weed-suppressive 'indica' rice line and a non-weed suppressive 'japonica' cultivar was used to study the genetic bases of variation for seedling germination, shoot length and dry matter weight. Thirteen QTL were detected and each QTL explained 5–10% of the phenotypic variation of the traits [76].

GWAS has been employed to investigate the genetic architecture of weed competition in *A. thaliana*, and rice [51, 55]. For example, a set of 195 accession of *A. thaliana* grown with the presence and absence of bluegrass, *Poa annua*, were analysed for trait (29 phenotypes related to phenology, resource acquisition, hoot architecture, seed dispersal, fecundity, reproductive strategy and survival)-marker association [51]. Several significant SNP associations for yield (fruit number on basal branches) with and without weed competition were identified. This study further identified a candidate gene, *TSF (TWIN SISTER OF FT)* which was associated with flowering time, duration of flowering, climate variation, the number of primary branches and escape strategy to competition, suggesting adaptive strategy to escape competition. However, no such study has been conducted in canola to identify genes which control weed competition and/or allelopathy.

3.3. Genetic variation for allelopathy

Allelopathy is a mechanism whereby a plant ensures itself a competitive advantage by placing phytotoxins into the adjacent environment [17]. Numerous allelochemicals that affect weed species have been identified and characterised [77]. Their existence varies with species and variety, and will almost always operate as a 'cocktail' of chemicals from any one source. An et al. [78], for example, showed that the allelopathic capability of *Vulpia* spp. involved more than 20 separate compounds. The role of allelopathy in suppression of weed growth has been studied in a range of crops including wheat [57], rice [79–82], barley [83], cotton [84], and sorghum [85].

Different laboratory based assays used to measure the allelopathy activity have been reviewed by Wu et al. [90]. These include the 'plant-box method' [86], the 'relay-seeding technique' [87], the 'equal-compartment-agar-method' or ECAM [88–90], and hydroponic methods [91, 92]. Generally, these assays involve growing of seedlings of the donor plants (*e.g.* crop species) in the presence of, or followed by, weed species for a short period of time. The allelopathic crops such as *Brassica rapa*, *B. juncea*, *B. nigra*, *B. hirta* and *B. napus* exude phytotoxic compounds [93–97] which suppress the growth of the weed species depending on the tolerance of the receiver plants to the chemicals being exuded. In the field, it is necessary to recognise that there would be an exchange of allelopathic chemicals between crop and weed with the outcome determined by relative potency of the allelochemicals and the tolerances of the receiving plants to the chemicals received [98]. Allelopathic activity is measured as the reduction of weed root growth in the presence of allelochemicals relative to that in the absence of the donor plants.

One question often raised is whether the laboratory method reflects performance under field conditions. Seal et al. [99] for rice and Asaduzzaman et al. [88] for canola both showed high correlations between the ECAM method in the laboratory and field performance. The other question is how field performance can be attributed to allelopathy. Unfortunately, there is no simple measure. In some cases inspection of the roots of affected plants show symptoms of inhibited development, such as root pruning, thickened roots and distortions not normally seen. In most cases, it has to be assumed that if field performance matches that in the laboratory then allelopathy is at least part of the explanation. Root exudates can be collected and analysed for bioactive compounds. Such compounds can be then applied to the receiver plants to ensure that the same outcome is achieved as described in [100]. Weidenhamer [101] has shown that it is possible to measure the presence of allelochemicals *in situ* in the rhizosphere using a sorptive coated stir bar inserted into the measurement zone for subsequent analysis by HPLC.

Phytotoxic allelochemicals have also been identified in *Brassica* plant residues and exudates that are known to suppress weed infestation [19, 95, 102]. *Brassica* species are also well known to synthesise glucosinolates which have shown allelopathic effects on pathogens due to the production of isothiocyanates. This process has been coined biofumigation [103, 104].

Genetic variation for allelopathy in canola and its related species, *Sinapis alba* L. has been studied [93, 105, 106]. Asaduzzaman et al. [107] investigated allelopathy among 70 diverse accessions of canola using annual ryegrass (*Lolium rigidum*) as the 'test' weed. The range of allelopathic impacts is shown in **Figure 3**. One *B. napus* cultivar of Australian origin, cv 'Av-Opal', was strongly allelopathic both in the laboratory and in the field whereas commercial cv. Barossa was at the other extreme in both laboratory and field. Field study showed that the allelopathic trait is independent of plant biomass and grain yield, and no consistent relationship between plant height and weed competitive ability was found among genotypes.

The greater weed suppression ability of cv. Av-Opal was confirmed in a two-year field study against annual ryegrass and other weeds (shepherd's purse, Indian hedge mustard and barley

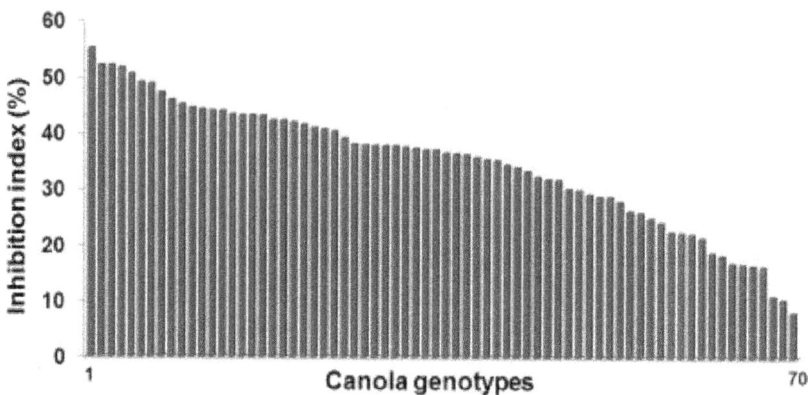

Figure 3. Allelopathic effect of 70 canola genotypes on root length of annual ryegrass seedlings (lsd = 10) [107].

grass) relative to cv. Barossa [28, 107]. Interestingly, Av-Opal was not exceptionally competitive as it is of short stature and poorly adapted to adverse environmental conditions [28]. In a subsequent study, Asaduzzaman et al. [108] investigated the biochemical basis of the allelopathy and detected numerous bioactive secondary metabolites including sinapyl alcohol, *p*-hydroxybenzoic acid and 3,5,6,7,8-pentahydroxy flavones in the root exudates. A comparison of the allelopathic capabilities between cv. Av-Opal and cv. Barossa is shown in **Figure 4**.

3.4. Detection of QTL for allelopathy

The genetic bases of allelopathy activity have been investigated in wheat [57, 110] and in rice [111–115]. For wheat, doubled haploid lines were developed from the strongly-allelopathic cultivar Tasman and the non-allelopathic cultivar Sunco. Significant differences were recorded for root growth of annual ryegrass between the doubled haploid lines [89]. Analysis of RFLP, AFLP and SSRs markers identified two major QTLs on chromosome 2B associated with wheat allelopathy.

In rice, several QTL have been detected across the rice genome and these QTL explain 5–36.6% of phenotypic variation in crop interference traits (**Table 1**). Jensen et al. [113] identified four major QTL on chromosomes 2, 3 and 8 which accounted for 35% of total variation of the allelopathic activity in the RIL population derived from japonica cv. IAC165 (allelopathic parent) and indica cv CO39. Ebana et al. [116] identified a major QTL on chromosome 6 accounting for 16.1% of the phenotypic variance in an F_2 population of 192 lines from indica line PI312777/japonica line Rexmont. Jensen et al. [114] identified QTL for RLSWRL and GHWRL on the same genomic marker interval, confirming that major genes for weed root length may be located in this region. The most important QTL were on chromosomes 3, 5, 8

Figure 4. A comparison of a strongly allelopathic cultivar (AV-opal, left) and a weakly allelopathic cultivar (Barossa, right) [109]. Barossa plot showing extensive growth of different weeds.

and 11 [111, 116]. This indicates that allelopathy activity in cereal is controlled by quantitative loci. The relatively low phenotypic variation for the individual QTL is explained by the difficulty in measuring the allelopathic traits at the individual genotype level.

4. Conclusions

Herbicide resistance is a major impediment in sustaining high crop productivity. The lack of new chemical modes of action becoming available emphasises the need for novel approaches to control weeds. Crop competitiveness and allelopathy are potential tools to reduce the dependence on synthetic chemical inputs and in so doing may extend the lives of key herbicides. A challenge for researchers is to be able to separate competitiveness from allelopathy in the field. For crop producers it does not really matter whether it is one or the other or both as long it works. A further challenge for researchers is attracting funds to undertake this work to commercial outcomes.

What are the prospects of herbicide resistance evolution occurring to allelochemicals? Of course the risks exist but they are likely to be much lower for at least two reasons: firstly allelopathy relies on a mix of chemicals at any one time from a single crop; and different crops have different mixes of chemicals so that in a rotation of crops, weeds will be exposed to chemicals of different modes of action only once or twice in a rotation cycle.

Phenotyping traits associated with allelopathic activity, such as reduction of weed growth in the laboratory and field, with high-throughput genotyping technology such as sequencing and mapping populations, allow researchers to detect QTL and genes associated with allelopathy and weed competition. It is an open question whether weed competition and allelopathy are distinct traits, but if this is the case, both traits could be pyramided in a single variety. In addition to genetic and phenotypic information, functional 'omic' data, such as identification of secondary metabolites, can be integrated in the QTL analysis leading to the detection of genes and pathways responsible for allelopathy activity. This would enable the development of novel alleles to expand breeding activity for weed interference in canola.

Acknowledgements

Authors thank NSW Department of Primary Industries, Charles Sturt University and E.H. Graham Centre for Agricultural Innovation for supporting crop interference research.

Conflict of interest

Authors do not have conflict of interest to declare.

Author details

Harsh Raman[1,2]*, Nawar Shamaya[1,2] and James Pratley[2]

*Address all correspondence to: harsh.raman@dpi.nsw.gov.au

1 NSW Department of Primary Industries, Wagga Wagga Agricultural Institute, Wagga Wagga, NSW, Australia

2 Graham Centre for Agricultural Innovation, Charles Sturt University, Wagga Wagga, NSW, Australia

References

[1] Nagaharu U. Genome analysis in Brassica with special reference to the experimental formation of *B. napus* and peculiar mode of fertilization. Journal of Japanese Botany. 1935;**7**:389-452

[2] Yang Y-W, Lai K-N, Tai P-Y, Li W-H. Rates of nucleotide substitution in angiosperm mitochondrial DNA sequences and dates of divergence between *Brassica* and other angiosperm lineages. Journal of Molecular Evolution. 1999;**48**(5):597-604

[3] Gómez-Campo C, Prakash S. Origin and domestication. In: Gómez-Campo C, editor. Biology of Brassica Coenospecies. Netherlands: Elsevier; 1999. pp. 33-58

[4] Raymer PL. Canola: An emerging oilseed crop. In: Janick J, Whipkey A, editors. Trends in New Crops and New Uses. Alexandria, VA: ASHS Press; 2002. pp. 122-126

[5] Colton B, Potter T, editors. History: The Organising Commitee of the 10th International Rapeseed Congress1999. pp. 1-4

[6] AOF. Australian Oilseeds Federation. Crop Report. AOF July 2017 http://www.australia-noilseeds.com/__data/assets/pdf_file/0012/11190/AOF_Crop_Report_July_2017.pdf. 2017

[7] Kirkegaard JA, Sprague SJ, Dove H, Kelman WM, Marcroft SJ, Lieschke A, et al. Dual-purpose canola - a new opportunity in mixed farming systems. Australian Journal of Agricultural Research. 2008;**59**(4):291-302

[8] Blackshaw RE, Lemerle D, Mailer R, Young KR. Influence of wild radish on yield and quality of canola. Weed Science. 2002;**50**:344-349. DOI: 101614/0043-1745

[9] Brooke G, McMaster C. Weed Control in Winter Crops. NSW Department of Primary Industries, Orange Australia ISSN 0812-907X. 2015

[10] OGTR Office of the Gene technology Regulator. The Biology of *Brassica napus* L. (Canola) and *Brassica juncea* (L.) Czern. &Coss. (Indian Mustard). Australian Government Department of Health; 2017. http:/www.ogtr.gov.au

[11] Pratley J, Broster J, Stanton R. Weed management. In: Principles of Field Crop Production. Wagga Wagga, www.csu.edu.au/__data/assets/pdf_file/0006/2805567/Chapter9_PratleyBrosterStanton.pdf: Graham Centre for Agricultural Innovation, Charles Sturt University; 2018

[12] Heap I. The International Survey of Herbicide Resistant Weeds – Online. www.weedscience.org [Accessed: 28 July 2017]; 2017

[13] Gressel J, Gassman AJ, Owen MDK. How well will stacked transgenic pest/herbicide resistances delay pests from evolving resistance? Pest Management Science. 2017;**73**:22-34

[14] Lemerle DV, Verbeek B, Coombes NE. Losses in grain yield of winter crops from *Lolium rigidum* (gaud.) depend on crop species, cultivar and season. Weed Research. 1995;**35**: 503-509

[15] Seavers GP, Wright KJ. Crop canopy development and structure influence weed suppression. Weed Research. 1999;**39**(4):319-328

[16] Donald CM. Competition among crop and pasture plants. In: Norman AG, editor. Advances in Agronomy. Vol. 15. New York, London: Academic Press; 1963. pp. 1-118

[17] Pratley JE. Allelopathy in annual grasses. Plant Protection Quartely. 1996;**11**:213-214

[18] Olofsdotter N, Rebulanan S. Weed-suppressing rice cultivars – Does allelopathy play a role? Weed Research. 1999;**39**(6):441-454

[19] Olofsdotter M, Jensen L, Courtois B. Improving crop competitive ability using allelopathy—An example from rice. Plant Breeding. 2002;**121**:1-9

[20] Worthington M, Reberg-Horton C. Breeding cereal crops for enhanced weed suppression: Optimizing allelopathy and competitive ability. Journal of Chemical Ecology. 2013; **39**:213-231

[21] Pierik R, Mommer L, Voesnek LA. Molecular mechanimsm of plant competition: Neighbour detection and response strategies. Functional Ecology. 2013;**27**:841-853

[22] Franklin KA. Shade avoidance. The New Phytologist. 2008;**179**:930-944

[23] Blackshaw RE. Differential competitive ability of winter wheat cultivars against downy brome. Agronomy Journal. 1994;**86**:649-654

[24] Morgan DC, Smith H. Linear relationship between phytochrome photoequilibrium and growth in plants under simulated natural irradiation. Nature. 1976;**262**:210-212

[25] Goodall J, Witkowski E, Ammann S, Reinhardt C. Does Allelopathy Explain the Invasiveness of *Campuloclinium macrocephalum* in the South African grassland Biological Invasion. 2010;**12**:3497-3512

[26] Challaiah OC, Burnside WGA, Johnson VA. Competition between winter wheat (*Triticum aestivum*) cultivars and downy brome (*Bromus tectorum*). Weed Science. 1986; **34**(5):689-693

[27] Lemerle D, Verbeek B, Cousens RD, Coombes NE. The potential for selecting wheat varieties strongly competitive against weeds. Weed Research. 1996;**36**(6):505-513

[28] Asaduzzaman M, Luckett D, Cowley R, An M, Pratley J. Canola cultivar performance in weed-infested field plots confirms allelopathy ranking from *in vitro* testing. Biocontrol Science and Technology. 2014;**24**:1394-1411

[29] Coleman R, Gill G, Rebetzke G. Identification of quantitative trait loci for traits conferring weed competitiveness in wheat (*Triticum aestivum* L.). Australian Journal of Agricultural Research. 2001;**52**:1235-1246

[30] Mokhtari S, Galwey N, Cousens R, Thurling N. The genetic basis of variation among wheat F_3 lines in tolerance to competition by ryegrass (*Lolium rigidum*). Euphytica. 2002;**124**:355-364

[31] Zhao DL, Atlin GN, Bastiaans L, Spiertz JHJ. Cultivar weed-competitiveness in aerobic rice: Heritability, correlated traits, and the potential for indirect selection in weed-free environments. Crop Science. 2006;**46**:372-380

[32] Zhao G, Atlin N, Bastiaans L, Spiertz J. Developing selection protocols for weed competitiveness in aerobic rice. Field Crops Research. 2006;**97**:272-285

[33] Asaduzzaman M, Pratley JE, Min A, Luckett DJ, Lemerle D. Canola interference for weed control. Springer Science Reviews. 2014;**2**:63-74

[34] Beckie H, Johnson E, Blackshaw R, Gan Y. Weed suppression by canola and mustard cultivars. Weed Technology. 2008;**22**:182-185

[35] Daugovish O, Thill D, Shafii B. Competition between wild oat (*Avena fatua*) and yellow mustard (*Sinapis alba*) or canola (*Brassica napus*). Weed Science. 2002;**50**:587-594

[36] Harker N, O'Donovan J, Blackshaw R, Johnson E, Holm F, Clayton G. Environmental effects on the relative competitive ability of canola and small-grain cereals in a direct-seeded system. Weed Science. 2011;**59**:404-415

[37] Shamaya N, Raman H, Rohan M, Pratley J, Wu H. Natural variation for interference traits against annual ryegrass in canola. Proceedings of AusCanola Conference, Perth; 2018

[38] Wenzl P, Carling J, Kudrna D, Jaccoud D, Huttner E, Kleinhofs A, et al. Diversity arrays technology (DArT) for whole-genome profiling of barley. Proceedings of National Academy of Sciences of the USA. 2004;**101**:9915-9920

[39] Wenzl P, Li H, Carling J, Zhou M, Raman H, Paul E, et al. A high-density consensus map of barley linking DArT markers to SSR, RFLP and STS loci and agricultural traits. BMC Genomics. 2006;**7**:206. https://doi.org/10.1186/1471-2164-7-206

[40] Williams J, Kubelik A, Livak K, Rafalski J, Tingey S. DNA polymorphisms amplified by arbitrary primers are useful as genetic markers. Nucleic Acids Research. 1990;**18**:6531-6535

[41] Vos P, Hogers R, Bleeker M, Reijans M, van de Lee T, Hornes M, et al. AFLP: A new technique for DNA fingerprinting. Nucleic Acids Research. 1995;**23**(21):4407-4414

[42] Lander ES, Botstein D. Mapping Mendelian factors underlying quantitative traits using RFLP linkage maps. Genetics. 1989;**121**(1):185-199

[43] Röder MS, Korzun V, Wendehake K, Plaschke J, Tixier M, Leroy P, et al. A microsatellite map of wheat. Genetics. 1998;**149**:2007-2023

[44] Li G, Quiros CF. Sequence-related amplified polymorphism (SRAP), a new marker system based on a simple PCR reaction: Its application to mapping and gene tagging in Brassica. Theoretical and Applied Genetics. 2001;**103**(2):455-461

[45] Baird NA, Etter PD, Atwood TS, Currey MC, Shiver AL, Lewis ZA, et al. Rapid SNP discovery and genetic mapping using sequenced RAD markers. PLoS One. 2008;**3**(10): e3376

[46] Trick M, Long Y, Meng J, Bancroft I. Single nucleotide polymorphism (SNP) discovery in the polyploid *Brassica napus* using Solexa transcriptome sequencing. Plant Biotechnology Journal. 2009;**7**:334-346

[47] Ganal MW, Altmann T, Röder MS. SNP identification in crop plants. Current Opinion in Plant Biology. 2009;**12**(2):211-217

[48] Gupta PK, Rustgi S, Mir RR. Array-based high-throughput DNA markers for crop improvement. Heredity. 2008;**101**(1):5-18

[49] Raman H, Raman R, Kilian A, Detering F, Carling J, Coombes N, et al. Genome-wide delineation of natural variation for pod shatter resistance in *Brassica napus*. PLoS One. 2014;**9**(7):e101673

[50] Poland JA, Brown PJ, Sorrells ME, Jannink J-L. Development of high-density genetic maps for barley and wheat using a novel two-enzyme genotyping-by-sequencing approach. PLoS One. 2012;**7**(2):e32253

[51] Frachon L, Libourel C, Villoutreix R, et al. Intermediate degrees of synergistic pleiotropy drive adaptive evolution in ecological time. Nat Ecology and Evolution. 2017;**1**:1551-1561

[52] Bartoli C, Roux F. Genome-wide association studies in plant pathosystems: Toward an ecological genomics approach. Frontiers in Plant Science. 2017;**8**:763. DOI: 10.3389/fpls.2017. 00763

[53] Botto JF, Coluccio MP. Seasonal and plant-density dependency for quantitative trait loci affecting flowering time in multiple populations of *Arabidopsis thaliana*. Plant, Cell and Environment. 2007;**30**(11):1465-1479

[54] Mutic JJ, Wolf JB. Indirect genetic effects from ecological interactions in *Arabidopsis thaliana*. Molecular Ecology. 2007;**16**(11):2371-2381

[55] Baron E, Richirt J, Villoutreix R, Amsellem L, Roux F. The genetics of intra- and interspecific competitive response and effect in a local population of an annual plant species. Functional Ecology. 2015;**29**(10):1361-1370

[56] Kikuchi S, Bheemanahalli R, Jagadish KSV, et al. Genome-wide association mapping for phenotypic plasticity in rice. Plant, Cell and Environment. 2017;**40**(8):1565-1575

[57] Wu H, Pratley J, Ma W, Haig T. Quantitative trait loci and molecular markers associated with wheat allelopathy. Theoretical and Applied Genetics. 2003;**107**:1477-1481

[58] Raman R, Taylor B, Marcroft S, Eckermann P, Rehman A, Lindbeck K, et al. Genetic map construction and localisation of qualitative and quantitative loci for blackleg resistance in canola (*Brassica napus* l.). 17th Crucifer Genetics Workshop, 5–8 Sept, Saskatoon, Canada; 2010

[59] Luckett DJ, Cowley R, Moroni S, Raman H. Improving water-use efficiency and drought tolerance in canola - potential contribution from improved carbon isotope discrimination (CID). Proceedings of the 13th International Rapeseed Congress, Prague; 2011

[60] Hou J, Long Y, Raman H, Zou X, Wang J, Dai S, et al. A tourist-like MITE insertion in the upstream region of the *BnFLC.A10* gene is associated with vernalization requirement in rapeseed (*Brassica napus* L.). BMC Plant Biology. 2012;**12**(1):238

[61] Raman R, Taylor B, Lindbeck K, Coombes N, Barbulescu D, Salisbury P, et al. Molecular mapping and validation of *Rlm1* genes for resistance to *Leptosphaeria maculans* in canola (*Brassica napus* L). Crop & Pasture Science. 2012;**63**:1007-1017

[62] Raman R, Taylor B, Marcroft S, Stiller J, Eckermann P, Coombes N, et al. Molecular mapping of qualitative and quantitative loci for resistance to *Leptosphaeria maculans*; causing blackleg disease in canola (*Brassica napus* L.). Theoretical and Applied Genetics. 2012;**125**(2):405-418

[63] Tollenaere R, Hayward A, Dalton-Morgan J, Campbell E, Lee JRM, Lorenc M, et al. Identification and characterization of candidate *Rlm4* blackleg resistance genes in *Brassica napus* using next-generation sequencing. Plant Biotechnology Journal. 2012;**10**(6):709-715

[64] Zou X, Suppanz I, Raman H, Hou J, Wang J, Long Y, et al. Comparative analysis of *FLC* homologues in Brassicaceae provides insight into their role in the evolution of oilseed rape. PLoS One. 2012;**7**(9):e45751

[65] Raman H, Raman R, Eckermann P, Coombes N, Manoli S, Zou X, et al. Genetic and physical mapping of flowering time loci in canola (*Brassica napus* L.). Theoretical and Applied Genetics. 2013;**126**:119-132

[66] Raman H, Dalton-Morgan J, Diffey S, Raman R, Alamery S, Edwards D, et al. SNP markers-based map construction and genome-wide linkage analysis in *Brassica napus*. Plant Biotechnology Journal. 2014;**12**(7):851-860

[67] Raman H, Raman R, Luckett D, Cowley R, Diffey S, Leah D, et al. Understanding the genetic bases of phenotypic variation in drought tolerance related traits in canola (*Brassica napus* L.) Proceedings of the 18th Australian Research assembly on Brassicas, 29th September-2nd October, 2014. p 119-125

[68] Raman R, Tanaka E, Coombes N, Diffey S, Lindbeck K, Price A, et al. Genome-wide association analyses identify novel loci for blackleg resistance in *Brassica napus*. 2015. https://event-wizardcom/files/clients/RKYES4VI/IRC2015_ABSTRACTS_July2015-webpdf. p. 69

[69] Larkan NJ, Raman H, Lydiate DJ, Robinson SJ, Yu F, Barbulescu DM, et al. Multi-environment QTL studies suggest a role for cysteine-rich protein kinase genes in quantitative resistance to blackleg disease in *Brassica napus*. BMC Plant Biology. 2016;**16**(1):1-16

[70] Liu J, Wang J, Wang H, Wang W, Zhou R, Mei D, et al. Multigenic control of pod shattering resistance in chinese rapeseed germplasm revealed by genome-wide association and linkage analyses. Frontiers in Plant Science. 2016;**7**:1058. DOI: 10.3389/fpls.2016.01058. e

[71] Raman H, Raman R, Coombes N, Song J, Diffey S, Kilian A, et al. Genome-wide association study identifies new loci for resistance to *Leptosphaeria maculans* in canola. Frontiers in Plant Science. 2016;**7**:1513. DOI: 10.3389/fpls.2016.01513

[72] Raman H, Raman R, Coombes N, Song J, Prangnell R, Bandaranayake C, et al. Genome-wide association analyses reveal complex genetic architecture underlying natural variation for flowering time in canola. Plant, Cell & Environment. 2016;**39**(6):1228-1239

[73] Raman R, Diffey S, Carling J, Cowley R, Kilian A, Luckett D, et al. Quantitative genetic analysis of yield in an Australian *Brassica napus* doubled haploid population. Crop & Pasture Science. 2016;**67**(4):298-307

[74] Raman H, Raman R, McVittie B, Orchard B, Qiu Y, Delourme R. A major locus for manganese tolerance maps on chromosome A09 in a doubled haploid population of *Brassica napus* L. Frontiers in Plant Science. 2017;**8**:1952. DOI: 10.3389/fpls.2017.01952

[75] Reid T. The Genetics of Competitive Ability in Spring Wheat. Alberta, Canada: University of Alberta; 2010

[76] Zhang H, Yu T, Huang Z, Zhu G. Mapping quantitative trait loci (QTLs) for seedling-vigor using recombinant inbred lines of rice. Field Crops Research. 2005;**91**:161-170

[77] Lankau R. A chemical trait creates a genetic trade-off between intra- and interspecific competitive ability. Ecology. 2008;**89**(5):1181-1187

[78] An M, Haig T, Pratley JE. Phytotoxicity of vulpia residues: II. Separation, identification, and quantitation of allelochemicals from Vulpia myuros. Journal of Chemical Ecology. 2000;**26**:1465-1476

[79] Kong CH, Chen XH, Hu F, Zhang SZ. Breeding of commercially acceptable allelopathic rice cultivars in China. Pest Management Science. 2011;**67**:1100-1106

[80] Dilday RH, Lin J, Yan W. Identification of allelopathy in the USDA-ARS rice germplasm collection. Australian Journal of Experimental Agriculture. 1994;**34**:907-910

[81] Gealy DR, Wailes EJ, Estorninos LE, Chavez RSC. Rice cultivar differences in suppression of barnyardgrass (*Echinochloa crus-galli*) and economics of reduced propanil rates. Weed Science. 2003;**51**:601-609

[82] Gealy DR, Yan W. Weed suppression potential of 'rondo' and other Indica Rice Germplasm lines. Weed Technology. 2012;**26**(3):517-524

[83] Liu DL, Lovett JV. Biologically active secondary metabolites of barley. II. Phytotoxicity of barley allelochemicals. Journal of Chemical Ecology. 1993;**19**(10):2231-2244

[84] Uludag A, Uremis I, Arslan M, Gozcu D. Allelopathy studies in weed science in Turkey – A review. Journal of Plant Diseases and Protection. 2006;**20**:419-426

[85] Einhellig FA, Souza IF. Phytotoxicity of sorgoleone found in grain Sorghum root exudates. Journal of Chemical Ecology. 1992;**18**(1):1-11

[86] Fujii Y. The potential biological control of paddy weeds with allelopathy-allelopathic effect of some rice varietie. In: Interantional Symposium Biological Control and Intreagted Management of Paddy and Aquatic Weeds in Asia. Tsukuba: National Agricultural Research Centre of Japan; 1992

[87] Navarez D, Olofsdotter M. Allelopathic rice for *Echinochloa crus-galli* control. In: Brown H, Cussans GW, Devine MDD, Fernandez-Quintanilla CSO, Helweg A, Labrada RE, Landes M, Kudsk PS, editors. 2nd International Weed Control Congress. Denmark; 1996

[88] Asaduzzaman M, An M, Pratley JE, Luckett DJ, Lemerle D. Laboratory bioassay for canola (*Brassica napus*) allelopathy. Journal of Crop Science and Biotechnology. 2014; **17**(4):267-272

[89] Wu H, Pratley JE, Lemerle D, Haig T. Laboratory screening for allelopathic potential of wheat (*Triticum aestivum*) accessions against annual ryegrass (*Lolium rigidum*). Australian Journal of Agricultural Research. 2000;**51**:259-266

[90] Wu H, Pratley JE, Lemerle D, An M, Liu DL. Autotoxicity of wheat (*Triticum aestivum* L.) as determined by laboratory bioassays. Plant and Soil. 2007;**296**:85-93

[91] Belz RG, Hurle K. Dose-response-a challenge for allelopathy? Nonlinearity. 2005;**3**:173-211

[92] Kim S, Madrid A, Park S, Yang S, Olofsdotter M. Evaluation of rice allelopathy in hydroponics. Weed Research. 2005;**45**:74-79

[93] Brown PD, Morra MJ. Hydrolysis products of glucosinolates in *Brassica napus* tissues as inhibitors of seed germination. Plant and Soil. 1996;**181**(2):307-316

[94] Buchanan AL, Kolb LN, Hooks CRR. Can winter cover crops influence weed density and diversity in a reduced tillage vegetable system? Crop Protection. 2016;**90**:9-16

[95] Haramoto ER, Gallandt ER. Brassica cover cropping: I. Effects on weed and crop establishment. Weed Science. 2005;**53**(5):695-701

[96] Petersen J, Belz R, Walker F, Hurle K. Weed suppression by release of isothiocyanates from turnip-rape mulch. Agronomy Journal. 2001;**93**(1):37-43

[97] Vaughn SF, Boydston RA. Volatile allelochemicals released by crucifer green manures. Journal of Chemical Ecology. 1997;**23**(9):2107-2116

[98] Moore JR, Asaduzzaman M, Pratley JE. Dual direction allelopathy: the case of canola, wheat and annual ryegrass Building Productive, Diverse and Sustainable Landscapes: Proceedings of the 17th Australian Society of Agronomy Conference, 20–24 September 2015. Hobart, Australia; 2015

[99] Seal AN, Pratley JE, Haig T. Can results from a laboratory bioassay be used as an indicator of field performance of rice cultivars with allelopathic potential against *Damasonium minus* (starfruit). Australian Journal of Agricultural Research. 2008;**59**:183-188

[100] Haig T, Haig TJ, Seal AN, Pratley JE, An M, Wu H. Lavender as a source of novel plant compounds for the development of a natural herbicide. Journal of Chemical Ecology. 2009;**35**:1129-1136

[101] Weidenhamer JD. Biomimetic measurement of allelochemical dynamics in the rhizosphere. Journal of Chemical Ecology. 2005;**31**:221-236

[102] Alcántara C, Pujadas A, Saavedra M. Management of *Sinapis alba* subsp. *mairei* winter cover crop residues for summer weed control in southern Spain. Crop Protection. 2011; **30**(9):1239-1244

[103] Kirkegaard JA, Sarwar M. Biofumigation potential of brassicas. Plant and Soil. 1998;**201**: 71-89

[104] Kirkegaard JA, Sarwar M. Glucosinolate profiles of Australian canola (*Brassica napus* L.) and Indian mustard (*Brassica juncea* L.) cultivars: Implications for biofumigation. Australian Journal of Agricultural Research. 1999;**50**(3):315-324

[105] Kruidhof HM, Bastiaans L, Kropff MJ. Cover crop residue management for optimizing weed control. Plant and Soil. 2009;**318**(1–2):169-184

[106] Boydston RA, Morra MJ, Borek V, Clayton L, Vaughn SF. Onion and weed response to mustard (*Sinapis alba*) seed meal. Weed Science. 2011;**59**(4):546-552

[107] Asaduzzaman M, An M, Pratley JE, Luckett DJ, Lemerle D. Canola (*Brassica napus*) germplasm shows variable allelopathic effects against annual ryegrass (*Lolium rigidum*). Plant and Soil. 2014;**380**:47-56

[108] Asaduzzaman M, Pratley JE, An M, Luckett DJ, Lemerle D. Metabolomics differentiattion of canola genotypes: Towards an understanding of canola alleolochemicals. Frontiers in Plant Science. 2015;**5**:765. DOI: 10.3389/fpls.2014.00765

[109] Asaduzzaman M, Luckett DJ, An M, Pratley JE, Lemerle D. Management of Paterson's curse (*Echium plantagineum*) through canola interference. In: Nineteenth Australasian Weeds Conference, Hobart; 2014. pp. 162-165

[110] Bertholdsson NO. Breeding spring wheat for improved allelopathic potential. Weed Research. 2010;**50**:49-57

[111] Zhou YJ, Cao CD, Zhuang JY, Zheng KL, Guo YQ, Ye M, et al. Mapping QTL associated with rice allelopathy using the rice recombinant inbred lines and specific secondary metabolite marking method. Allelopathy Journal. 2007;**19**:479-485

[112] Ebana K, Yan W, Dilday R, Namai H, Okuno K. Analysis of QTL associated with the allelopathic effect of rice using water-soluble extracts. Breeding Science. 2001;**51**:47-51

[113] Jensen LB, Cortois B, Shen LS, Li ZK, Olofsdotter M, Mauleon RP. Locating genes controlling allelopathic effects against barnyard grass in upland rice. Agronomy Journal. 2001;**93**:21-26

[114] Jensen LB, Courtois B, Olofsdotter M. Quantitative trait loci analysis of allelopathy in rice. Crop Science. 2008;**48**:1459-1469

[115] Chen XH, Hu F, Kong CH. Varietal improvement in rice allelopathy. Allelopathy Journal. 2008;**22**:379-384

[116] Ebana K, Yan W, Dilday RH, Namai H, Okuno K. Variation in allelopathic effect of rice (*Oryza sativa* L.) with water-soluble extracts. Agronomy Journal. 2001;**93**:12-16

[117] Navarez DC, Olofsdotter M, editors. Relay Seeding Technique for Screening Allelopathic Rice (*Oryza sativa*). Copenhagen, Denmark: The Second International Weed Control Congress; 1996

[118] Dali Z, Qian Q, Sheng T, Guojun D, Fujimoto H, Yasufumi K, et al. Genetic analysis of rice allelopathy. Chinese Science Bulletin. 2003;**48**:265-268

Potentially Phytotoxic of Chemical Compounds Present in Essential Oil for Invasive Plants Control: A Mini-Review

Mozaniel Santana de Oliveira,
Wanessa Almeida da Costa,
Priscila Nascimento Bezerra,
Antonio Pedro da Silva Souza Filho and
Raul Nunes de Carvalho Junior

Additional information is available at the end of the chapter

http://dx.doi.org/10.5772/intechopen.74346

Abstract

The control of invasive plants is still carried out with the use of synthetic chemical agents that may present high toxicity and, consequently, be harmful to humans and animals. In Brazil, especially in the Amazon, small producers use this kind of technique in a rustic way, with brushcutters or fire. In this sense, the search for natural agents with bioherbicide potential becomes necessary. Examples of these agents are the essential oils that over the years have been shown to be a viable alternative to weed control. Thus, this review aims to show the potentially phytotoxic activity of allelochemicals present in essential oils of different aromatic plants.

Keywords: natural products, essential oils, allelochemicals, allelopathy

1. Introduction

The performance of agricultural activity in tropical regions, both in fertile and in low fertility soils, has been limited by the occurrence of a series of extremely aggressive and diverse plants, called weeds. The main consequence of crop infestation by these plants is increasing costs to maintain the crops and reduction of productivity and its consequent competitive capacity. These plants may also represent an additional problem for farmers either because they are often toxic to different animals or because they are permanent sources for the spread of diseases to crop plants [1]. In this context, weed management and control become crucial both from the point of view of crop productivity and the profitability of the farming system.

In modern agriculture, where high yields are expected, in the face of increasing demands for food – due to the increasing world population – the control of these plants has been made, basically, by the use of chemical herbicides. However, such a procedure may not be sustainable over time, especially because it conflicts with the interests of modern society, which is increasingly concerned with the quality of food and with the preservation of natural resources. At the same time, the reduction in the efficiency of the current products available in the market has been observed as a consequence of the appearance of resistant plants [2, 3], leading to an increase in the use of herbicides or the contractions employed, which only increases the problem. All these factors point to the need of science to make available new and revolutionary methods of weed control.

A viable alternative to this challenge are the numerous chemically diverse compounds produced by plants that may offer new chemical structures capable of efficiently replace those already available in the market. In this line, crude extracts and isolated or associated chemical substances can be an excellent strategy to partially or totally replace the use of herbicides.

Over the last decades, different chemical compounds with bioherbicidal properties have been isolated and identified in different plants [4–7]. Among the many chemical classes with potential use in weed management, the secondary metabolites present in essential oils can be highlighted, since the different chemical classes of volatile compounds are notable for the wide potential of use in different activities of interest for humanity and specifically in the management of weeds.

2. Allelopathy history

Allelopathy is the chemical interaction between plants and other living organisms [8]. There are two types of interactions between plants: a phytotoxic one, which inhibits the germination of seeds and the development of the radicle and hypocotyl [9], and a stimulatory effect, which favors the development of the plant [10]. The chemical substances responsible for the allelopathic effect are called allelochemicals [11].

The allelopathy is a relatively new science, having its basic concepts established over the last 8 decades. However, chemical interactions among plants are not exactly new, since reports on the subject are found in old references. [12–16]. In the 1800s, several phenomena were attributed to the chemical interaction among plants [17]. In the early 1900's, [18] reported the presence of toxic compounds produced by plants that could be extracted from the soil. The first reports proving the interference promoted by chemical compounds were developed in the 1960's [19], showing that the volatile compounds were affecting the dynamics among plants.

3. Control of invasive plants

Currently, the chemical control method is the most used to inhibit the growth of invasive plants, which includes the use of synthetic herbicides, in large quantities, mainly by large producers, as reported by some authors [20, 21]. The use of synthetic and toxic chemical herbicides in management areas promotes the death of weeds in a selective way and, consequently, it ends the competition among the plants, helping to increase the production of green mass in the pasture [22]. The increasing use of agrochemicals may represent an unsustainable practice because these pesticides can pollute the environment and promote the contamination of various animal species. Also, new insecticide-resistant insects are appearing and invasive plants that are tolerant to modern herbicides are becoming more frequent [23].

Weed resistance to herbicides may be related to an evolutionary process; however, some developments of resistant weed biotypes are imposed by agriculture through selection pressure caused by the intensive use of herbicides. Weed resistance to herbicides may result from biochemical, physiological, morphological or phenological changes of certain invasive plant biotypes. Many cases of resistance to herbicides result from either the alteration of the site of action of the herbicide or the increase of its metabolism, or the departmentalization and compartmentalization of the herbicide in the plant [24, 25]. This way, allelopathy can be a natural alternative for the control of invasive plants.

4. Volatile allelochemicals

Weeds promote two basic types of interference in agricultural crops: allelospoly and allelopathy. Allelospoly is the type of interference promoted by competition for essential factors to the species survival, such as water, nutrients and physical space. Allelopathy involves the production of allelochemicals and subsequent release into the environment [26]. Almost all allelochemicals exist in conjugated, non-toxic forms. The toxic fragment can be released after exposure to stress or after tissue death [27].

The use of allelopathy for weed control may be an ecologically viable alternative [28]. Thus, the use of essential oils with phytotoxic potential is becoming widespread, since the allelochemicals present in these oils generally have low cytotoxicity. For example, [29] evaluated the effect of *Carum carvis* essential oils rich in carvone (71.08%) and limonene (25.42%), and verified that this oil has a strong phytotoxic activity on seed germination and radicle elongation of *Linum usitatissimum*, *Phalaris canariensis* and *Triticum aestivum*.

Another example is the eucalypt essential oil that has a rich chemical composition in 1,8-cineole (58.3%), α-pinene (17.3%) and α-thujene (15.5%), which significantly inhibited seed germination of *Sinapis arvensis*, *Diplotaxis harra* and *Trifolium campestre*, in different intensities according to the recipient species, demonstrating that each species has a different specificity. In addition, the application of post-emergence oil causes inhibition of chlorophyll production, leading to injuries such as chlorosis, necrosis and even complete wilting of plants [30].

Plant species such as *Origanum onites L.* and *Rosmarinus officinalis L.* also show strong alle-lopathic activity on species of *Poaceae* and invasive plants, by suppressing germination rate and elongation of radicle and hypocotyl [31]. The phytotoxic effects related to these two spe-cies of aromatic plants may be related to their rich chemical composition in the oxygenated monoterpenes 1,8-cineole, linalool, camphor and carvacrol and the monoterpene hydrocar-bon p-cymene [32–35], however, compounds found in lower concentrations as methyl phen-ylpropanoids have also demonstrated good allelopathic activity [36].

In the case of essential oils for the control of invasive plants, it is usually analyzed the effects of individual form, attributing the phytotoxic activity to only one component [37, 38]. However, the effects of volatile oils can also be related to the mixture of compounds, such as *Artemisia scoparia* oil which has a mixture of compounds such as monoterpene hydrocar-bons, oxygenated monoterpenes, sesquiterpene hydrocarbons, oxygenated sesquiterpenes, aliphatic compounds and other aromatic compounds [39]. The chemical composition of the essential oils depends on the biosynthetic path of the different classes of compounds, as can be observed in **Figure 1**, which brings the biosynthesis of some classes of volatile compounds.

Compounds such as eucalyptol, β-phellandrene, hexyl butanoate, *p*-cymene, α-ionone, (z)-3-oc-ten-1-ol, theaspirane a, vitispirane, dihydro-(−)-neoclovene, β-caryophyllene, (e)-2-octen-1-ol, a-terpineol, dehydro-ar-ionene, methyl salicylate, (z)-b-damascenone, (z)-dehydro-ar-ionene,

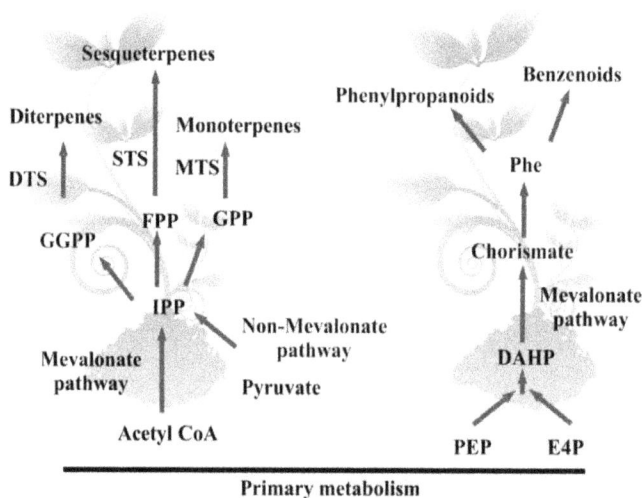

Figure 1. Biosynthesis of plant volatiles. Overview of biosynthetic pathways leading to the emission of plant volatile organic compounds. The plant precursors originate from primary metabolism. Abbreviations: DTS: Diterpene synthase; FPP: farnesyldiphosphate; GGPP: geranylgeranyldiphosphate; GLVs: green-leaf volatiles; GPP: geranyldiphosphate; IPP: isopentenyl pyrophosphate; MTS: Monoterpene synthase; STS: Sesquiterpene synthase; DAHP: 3-deoxy-D-arabinoheptulosonate-7 phosphate; E4P: erythrose 4-phosphate; PEP: phosphoenolpyruvate; Phe: phenylalanine. This flowchart was adapted from [40] and [41].

10-(tetrahydro-pyran-2-yloxy)-tricyclo[4.2.1(2,5)]decan-9-ol, (−)-caryophylleneoxide,dihydro-β-ionone, viridiflorol, cubenol, caryophyllene, α-bisabolol oxide-b, tetracosane and n-hexa-decane can be found in *Anisomeles indica* essential oil and also present good phytotoxic activity against invasive plants [42]. As well as *P. heyneanus Benth* essential oils, rich in patchouli alcohol, α-bulnesene, α-guaiene, seichelene and α-patchulene, and *P. hispidinervium* C. DC oils, rich in safrole, terpinolene, (E)-β-ocimene, δ-3-carene and pentadecane [43].

4.1. Monoterpenes

The monoterpenes have presented good phytotoxic activity, and reports of the use of these compounds to control plants refer to the 1960s [44]. This activity depends on the structural

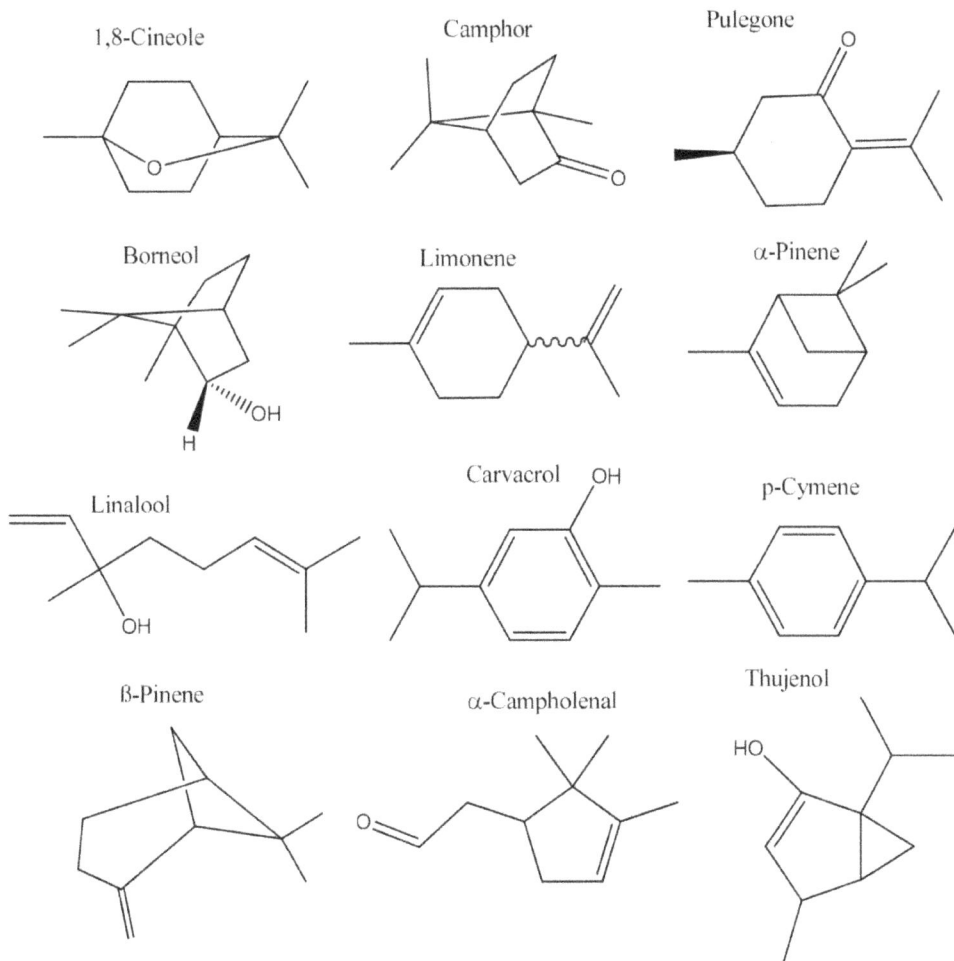

Figure 2. Chemical structures of oxygenated and non-oxygenated monoterpenes with bioherbicidal action.

characteristics of the molecules; for example, oxygenated monoterpenes exhibit different effects on germination and seedling development, and also alter cellular respiration, which impairs energetic metabolism [33, 34]. However, these phytotoxic effects promoted by a chemical species depend on its concentration, for example, *Lactuca sativa* essential oil composed essentially of α-pinene (16.00%), 1,8-cineole (66.93%) and pimonene (10.04%) presents different rates of germination inhibition [45].

In general, oxygenated monoterpenes have the highest phytotoxic effects over non-oxygenated [46]. However, there are non-oxygenated volatile molecules such as limonene which also have good phytotoxic activity [47]. Some monoterpenes had high inhibitory activity on germination and radicle elongation, and this may be related to the anatomical and physiological changes in the host plants, as well as to the reduction in some organelles such as mitochondria, and accumulation of lipid globules in the cytoplasm [48]. In **Figure 2**, the chemical structures of some monoterpenes with phytotoxic activity can be observed.

4.2. Sesquiterpenes

Bioassays have demonstrated that the sesquiterpenic allelochemicals β-cariofilene, β-copaene, spathulenol, germacrene B, bicyclogermacrene, globulol, viridiflorol, a-guaiene, and g-elemene have presented phytotoxity against various invasive plants and, in some cases, promote inhibition of other plants development, when they are close to species that produce these secondary metabolites [49–51]. Authors compared the effects of essential oils rich in sesquiterpenes and others rich in monoterpenes and found that the effects presented by sesquiterpenes, in some cases, may be smaller in relation to the affections exhibited by monoterpenes [52]. **Figure 3** shows the chemical structures of oxygenated and non-oxygenated sesquiterpenes with phytotoxic action.

However, this depends largely on the presence of oxygenated and non-oxygenated, cyclic or acyclic molecules, because depending on the molecular conformation the allelopathic effect may be higher or lower [53, 54]. This justifies the results obtained by other authors [55], who analyzed the effects of fractions of essential oils of *E. adenophorum,* of the inflorescence region, rich in sesquiterpenes, and its root rich in monterpenes. When the oils were tested at the same concentration (1 μL/mL), they inhibited germination and seedling elongation at the same ratio.

4.3. Phenylpropanoids

Phenylpropanoids are a class of secondary metabolites that are also naturally present in plants, and have exhibited strong phytotoxic activity against invasive plants. In 2016, [9] demonstrated that eugenol is the main active ingredient of clove essential oil and is also the agent possibly promoting phytotoxic activity against the invasive plants *Mimosa pudica* and *Senna obtusifolia.* Other authors also report the potentially allelopathic activity

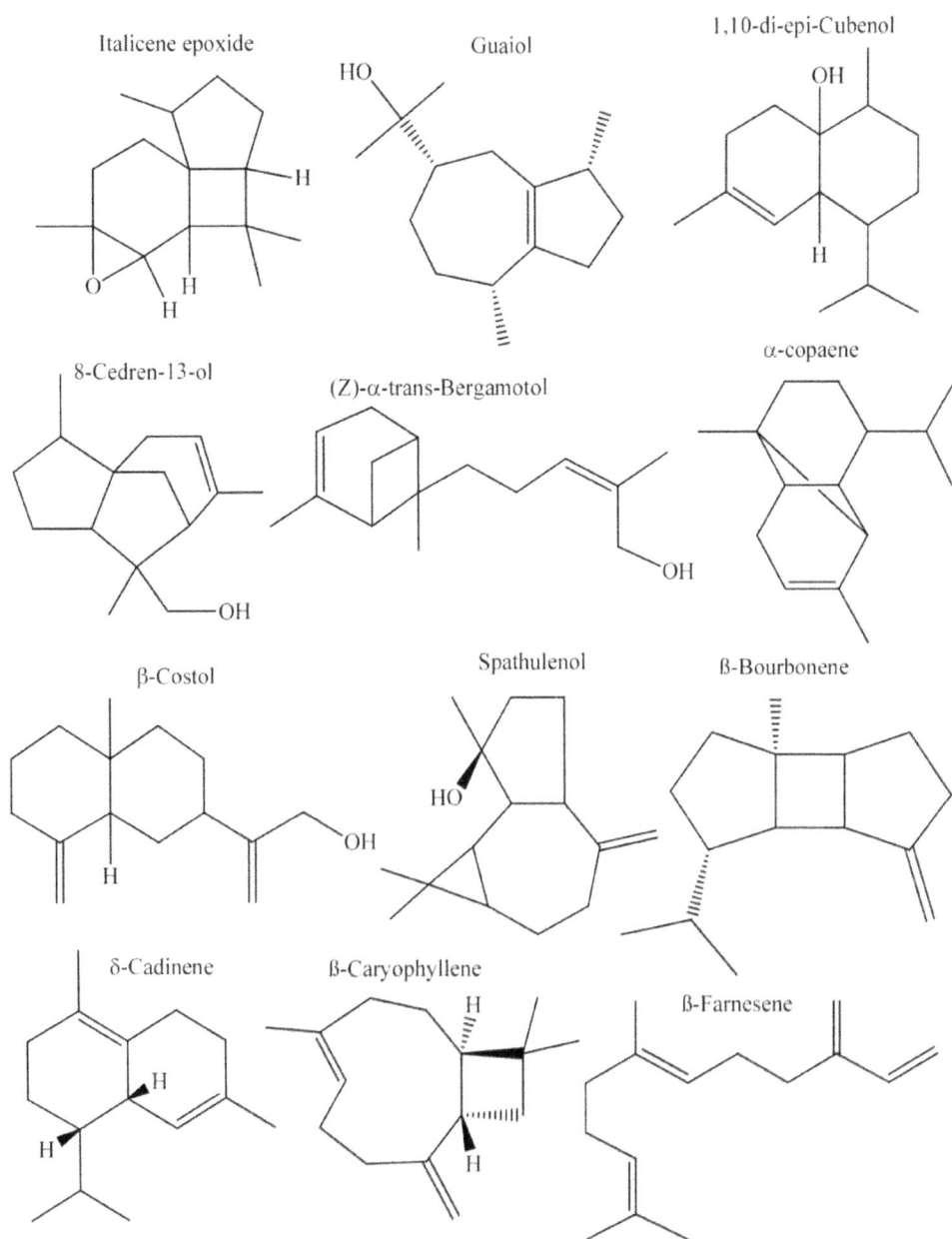

Figure 3. Chemical structures of oxygenated and non-oxygenated sesquiterpenes with bioherbicidal action.

of clove essential oil *Syzygium aromaticum* [56–58]. In addition to eugenol, other phenylpropanoids present in essential oils with phytotoxic activity are eugenyl acetate, safrole, methyl eugenol, anethole, myristicin, estragole, anethole and trans-anethole [36, 59–64]. **Figure 4**

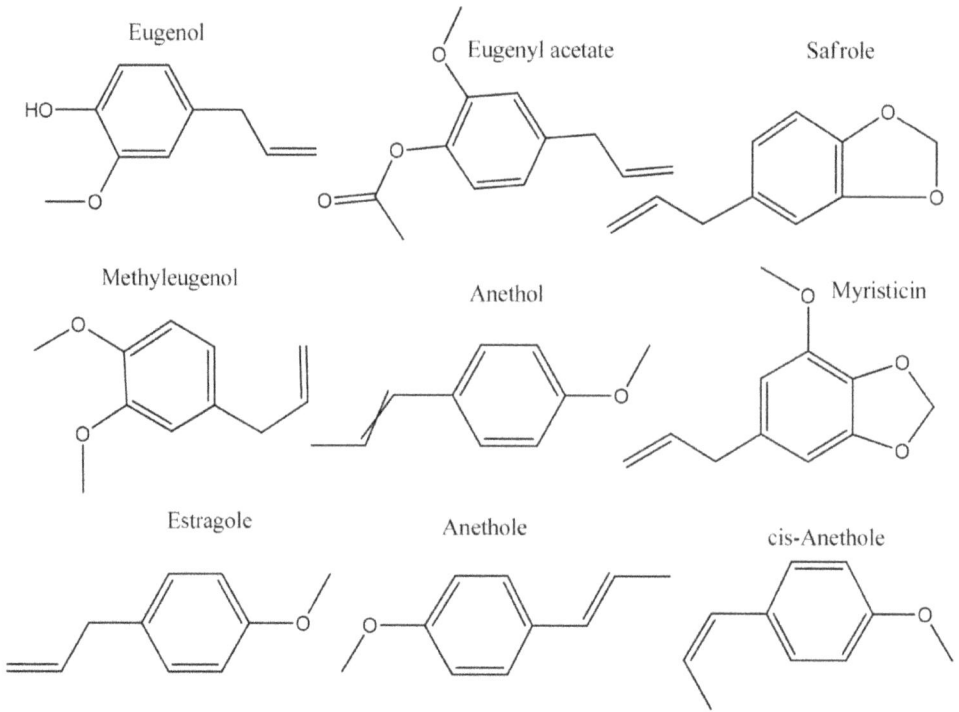

Figure 4. Chemical structures of phenylpropanoids with bioherbicidal action.

shows the chemical structures of the phenylpropanoids with potential use for control of invasive plants.

5. Conclusion

For essential oils to have good phytotoxic activity, some factors such as chemical composition, concentration and host plants may be taken into account. Among the monoterpene allelo-chemicals we can highlight the 1,8 cineole, among the sesquiterpenes or β-caryophyllene and among phenylpropanoids, eugenol. On the other hand, one of the difficulties that can appear for the use in large scale of essential oils is the volatility of their components.

Acknowledgements

Oliveira MS (Process Number: 1662230) and Costa WA (Process Number: 1427204) thank CAPES for the doctorate scholarship.

Author details

Mozaniel Santana de Oliveira[1]*, Wanessa Almeida da Costa[2], Priscila Nascimento Bezerra[1], Antonio Pedro da Silva Souza Filho[3] and Raul Nunes de Carvalho Junior[1,2]*

*Address all correspondence to: mozaniel.oliveira@yahoo.com.br and raulncj@ufpa.br

1 LABEX/FEA (Faculty of Food Engineering), Program of Post-Graduation in Food Science and Technology, Federal University of Para, Belém, Pará, Brazil

2 Program of Post-Graduation in Natural Resources Engineering, Federal University of Para, Belém, Pará, Brazil

3 Laboratory of Agro-industry, Embrapa Eastern Amazon, Belém, Pará, Brazil

References

[1] Poletti M, Omoto C. Resistência de Inimigos Naturais a Pesticidas. Rev Biotecnol Ciência e Desenvolv. 2003;**30**:1-17

[2] Norsworthy JK, Ward SM, Shaw DR, Llewellyn RS, Nichols RL, Webster TM, Bradley KW, Frisvold G, Powles SB, Burgos NR, Witt WW, Barrett M. Reducing the risks of herbicide resistance: Best management practices and recommendations. Weed Science [Internet]. Jan 20, 2012;**60**(SP1):31-62. DOI: http://www.bioone.org/doi/full/10.1614/WS-D-11-00155.1

[3] Devine MD, Shukla A. Altered target sites as a mechanism of herbicide resistance. Crop Protection [Internet]. Sep 2000;**19**(8-10):881-889. Available from: http://linkinghub.elsevier.com/retrieve/pii/S026121940000123X

[4] Souza Filho APS. In: da Silva Sousa Filho AP, editor. Ecologia química: A experiência brasileira. 1st ed. Belém, PA: Embrapa Amazônia Oriental, Belém; 2008. 150 p

[5] Santos JCF, Edilene GM, Marchi CS. In: de Miranda FVC, do Nascimento FE, de Oliveira JF, editors. Daninhas do Café. 1st ed. Planaltina, DF- Brazil: Empresa Brasileira de Pesquisa Agropecuária Embrapa Cerrados Ministério da Agricultura, Pecuária e Abastecimento; 2008

[6] Filho AJC, Santos LS, Guilhon GMSP, Moraes RPC, dos Santos RA, Filho AP da SS, Felizzola JF. Identified substances from the leaves of *Tephrosia cinerea* (Leguminoseae) crude extracts and their phytotoxic effects. International Journal of Life-Sciences Scientific Research [Internet]. Jul 6, 2017;**3**(4):1137-1141. Available from: http://ijlssr.com/currentissue/Tephrosia cinerea (Leguminoseae) Crude Leaves Extracts and their Phytotoxic Effects.pdf

[7] Pereira SG, Soares AM dos S, Guilhon G, Santos LS, Pacheco LC, Souza Filho AP da S. Phytotoxic potential of piperine and extracts from fruits of *Piper tuberculatum* Jaq. on Senna obtusifolia and *Mimosa pudica* plants. Allelopathy Journal 2016;**38**(1):91-102

[8]   Bostan C, Butnariu M, Butu M, Ortan A, Butu A, Rodinoc S, Parvue C. Allelopathic effect of *Festuca rubra* on perennial grasses. Romanian Biotechnology Letters. 2013;**18**(2):8190-8196

[9]   de Oliveira MS, da Costa WA, Pereira DS, Botelho JRS, de Alencar Menezes TO, de Aguiar Andrade EH, da Silva SHM, da Silva Sousa Filho AP, de Carvalho RN. Chemical composition and phytotoxic activity of clove (*Syzygium aromaticum*) essential oil obtained with supercritical CO_2. Journal of Supercritical Fluids [Internet]. 2016;**118**:185-193. DOI: http://dx.doi.org/10.1016/j.supflu.2016.08.010

[10]   Batista C de CR, de Oliveira MS, Araújo ME, Rodrigues AMC, Botelho JRS, da Silva Souza Filho AP, Machado NT, Carvalho RN. Supercritical CO2 extraction of açaí (*Euterpe oleracea*) berry oil: Global yield, fatty acids, allelopathic activities, and determination of phenolic and anthocyanins total compounds in the residual pulp. Journal of Supercrit Fluids [Internet]. Jan 2016;**107**:364-369. DOI: http://dx.doi.org/10.1016/j.supflu.2015.10.006

[11]   Latif S, Chiapusio G, Weston LA. Allelopathy and the role of allelochemicals in plant defence. In: Becard G, editor. Advances in Botanical Research [Internet]. 82nd ed. 2017;**82**: 19-54. Available from: https://www.sciencedirect.com/science/article/pii/S0065229616301203

[12]   Pavlychenko TK, Harrington J. Competitive efficiency of weeds and cereal crops. Canadian Journal of Research. 1934;**10**:77-94

[13]   Lucas CE. The ecological effects of external metabolites. Biological Reviews of the Cambridge Philosophical Society [Internet]. 1947;**22**(3):270-295. Available from: http://www.ncbi.nlm.nih.gov/pubmed/20253195

[14]   Rice EL. Allelopathy – An update. The Botanical Review. 1979;**45**(1):15-109

[15]   Rice EL. Allelopathy. In: Rice EL, editor. 2nd ed. New York: Academic Press; 1984. 422 p

[16]   Lee IK, Monsi M. Ecological studies on Pinus densiflora forest. Effect of plant substances in the floristic composition of undergrowth. Bot Manager. 1963;**76**:400-413

[17]   Stickney JS, Hoy PR. Toxic action of black walnut. Transactions of the Wisconsin State Horticultural Society. 1881;**11**:166-167

[18]   Schreiner O, Reed HS. The production of deleterious excretions by roots. Bull Torrey Bot Club [Internet]. 1907 Jun;**34**(6):279-303. Available from: http://www.jstor.org/stable/2479157?origin=crossref

[19]   Muller CH. Inhibitory terpenes volatilized from Salvia shrubs. Bulletin of the Torrey Botanical Club. 1965:38-45

[20]   Bueno MR, Alves GS, Paula ADM, Cunha JPAR. Volumes de calda e adjuvante no controle de plantas daninhas com glyphosate. Planta Daninha [Internet]. 2013;**31**(3): 705-713 Available from: http://www.scielo.br/scielo.php?script=sci_arttext&pid=S0100-83582013000300022&lng=pt&nrm=iso&tlng=en

[21]   Gazziero DLP. Misturas de agrotóxicos em tanque nas propriedades agrícolas do Brasil. Planta Daninha [Internet]. Mar 2015;**33**(1):83-92 Available from: http://www.scielo.br/scielo.php?script=sci_arttext&pid=S0100-83582015000100083&lng=pt&nrm=iso&tlng=en

[22] Ferreira E, Procópio S, Galon L, Franca A, Concenço G, Silva A, Aspiazu I, Silva A, Tironi S, Rocha PR. Manejo de plantas daninhas em cana-crua. Planta Daninha [Internet]. 2010 Dec;28(4):915-925 Available from: http://www.scielo.br/scielo.php?script= sci_arttext&pid=S0100-83582010000400025&lng=pt&tlng=pt

[23] Souza Filho AP, Alves S de M. Alelopa tia em ecossistema de pastagem cultivada [Internet]. 1st ed. Belém: Embrapa; 1998. 71 p. Available from: http://www.infoteca. cnptia.embrapa.br/handle/doc/388173

[24] Christoffoleti PJ, Victoria R, Filho S, Da CB. Resistência de planta s daninhas aos herbici-das. Planta Daninha. 1994;12(1990):13-20

[25] LeBaron HM, McFarland J. Herbicide resistance in weeds and crops. In: Society AC, editor. Herbicides – Current Research and Case Studies in Use [Internet]. InTech; 1990. pp. 336-352. Available from: http://www.intechopen.com/books/herbicides-current-research-and-case-studies-in-use/herbicide-resistant-weeds-the-technology-and-weed-management

[26] Spiassi A, Konopatzki MRS, Nóbrega LHP. Estratégias de manejo de plantas invaso-ras. Revista Varia Scientia Agrárias. [Internet]. 2010;1:177-188 Available from: http://e-revista.unioeste.br/index.php/variascientiaagraria/article/viewArticle/3493

[27] Putnam AR. Allelochemicals from plants as herbicides. Weed Science Society of America [Internet]. May 1, 1988;2(4):510-518 Available from: http://www.jstor.org/stable/3987390

[28] Macías FA, Molinillo JM, Varela RM, Galindo JC. Allelopathy – A natural alternative for weed control. Pest Management Science [Internet]. 2007 Apr;63(4):327-348. DOI: http:// doi.wiley.com/10.1002/ps.1342

[29] Marichali A, Hosni K, Dallali S, Ouerghemmi S, Bel Hadj Ltaief H, Benzarti S, Kerkeni A, Sebei AH. Allelopathic effects of *Carum carvi* L. essential oil on germination and seed-ling growth of wheat, maize, flax and canary grass. Allelopathy Journal [Internet]. 2014;34(1):81-94. Available from: http://search.proquest.com/openview/1742af81c669e6 bce54668881b849f28/1?pq-origsite=gscholar

[30] Grichi A, Nasr Z, Khouja ML. Phytotoxic effects of essential oil from *Eucalyptus lehm-anii* against weeds and its possible use as a bioherbicide. Bulletin of Environment, Pharmacology and Life Sciences [Internet]. 2016;5(March):17-23. Available from: http:// bepls.com/beplsmarch2016/4.pdf

[31] Atak M, Mavi K, Uremis I. Bio-Herbicidal Effects of oregano and rosemary essential oils on germination and seedling growth of bread wheat cultivars and weeds. Romanian Biotechnology Letters. 2015;21(1):11149-11159

[32] Romagni JG, Allen SN, Dayan FE. Allelopathic effects of volatile cineoles on two weedy plant species. Journal of Chemical Ecology. 2000;26(1):303-313

[33] Singh HP, Batish daizy R, Kaur S, Ramezani H, Kohli RK. Comparative phytotoxicity of four monoterpenes against *Cassia occidentalis*. Annals of Applied Biology [Internet]. Oct 2002;141(2):111-116. DOI: http://doi.wiley.com/10.1111/j.1744-7348.2002.tb00202.x

[34] Kordali S, Cakir A, Ozer H, Cakmakci R, Kesdek M, Mete E. Antifungal, phytotoxic and insecticidal properties of essential oil isolated from *Turkish Origanum* acutidens and its three components, carvacrol, thymol and p-cymene. Bioresource Technology [Internet]. Dec 2008;**99**(18):8788-8795. Available from: http://linkinghub.elsevier.com/retrieve/pii/S0960852408003696

[35] Jassbi AR, Zamanizadehnajari S, Baldwin IT. Phytotoxic volatiles in the roots and shoots of *Artemisia tridentata* as detected by headspace solid-phase microextraction and gas chromatographic-mass spectrometry analysis. Journal of Chemical Ecology. 2010;**36**(12):1398-1407

[36] De Almeida LFR, Frei F, Mancini E, De Martino L, De Feo V. Phytotoxic activities of Mediterranean essential oils. Molecules. 2010;**15**(6):4309-4323

[37] Singh HP, Batish DR, Kaur S, Kohli RK, Arora K. Phytotoxicity of the volatile monoterpene citronellal against some weeds. Zeitschrift fur Naturforsch – Section C Journal of Biosciences 2006;**61**(5-6):334-340

[38] Batish DR, Singh HP, Setia N, Kaur S, KRK. Chemical composition and phytotoxicity of volatile essential oil from intact and fallen leaves of *Eucalyptus citriodora*. Zeitschrift fur Naturforsch – Section C Journal of Biosciences. 2006;**61**(7-8):465-471

[39] Kaur S, Singh HP, Mittal S, Batish DR, Kohli RK. Phytotoxic effects of volatile oil from *Artemisia scoparia* against weeds and its possible use as a bioherbicide. Industrial Crops and Products [Internet]. 2010;**32**(1):54-61. DOI: http://dx.doi.org/10.1016/j.indcrop.2010.03.007

[40] Kant MR, Bleeker PM, Wijk M Van, Schuurink RC, Haring MA. Chapter 14: Plant volatiles in defence. In: Advances in Botanical Research [Internet]. 1st ed. Elsevier Ltd.; 2009. pp. 613-666. DOI: http://dx.doi.org/10.1016/S0065-2296(09)51014-2

[41] Dudareva N, Klempien A, Muhlemann JK, Kaplan I. Biosynthesis, function and metabolic engineering of plant volatile organic compounds. New Phytologist [Internet]. Apr 2013;**198**(1):16-32. DOI: http://doi.wiley.com/10.1111/nph.12145

[42] Batish DR, Singh HP, Kaur M, Kohli RK, Singh S. Chemical characterization and phytotoxicity of volatile essential oil from leaves of *Anisomeles indica* (Lamiaceae). Biochemical Systematics and Ecology [Internet]. Apr 2012;**41**:104-109. DOI: http://dx.doi.org/10.1016/j.bse.2011.12.017

[43] Souza Filho APDS, de Vasconcelos MAM, Zoghbi MDGB, Cunha RL. Efeitos potencialmente alelopáticos dos óleos essenciais de *Piper hispidinervium* C. DC. e *Pogostemon heyneanus* Benth sobre plantas daninhas. Acta Amazonica [Internet]. 2009;**39**(2):389-395. Available from: http://www.scielo.br/scielo.php?script=sci_arttext&pid=S0044-59672009000200018&lng=pt&tlng=pt

[44] Asplund RO. Some quantitative aspects of the phytotoxicity of monoterpenes. Weed Science Society of America. 1969;**17**(4):454-455

[45] de Oliveira CM, Cardoso M das G, Figueiredo AC da S, de Carvalho MLM, Miranda CASF de, Marques Albuquerque LR, Lee Nelson D, de Souza Gomes M, Silva LF, de Andrade Santiago J, Teixeira ML, Brandão RM. Chemical composition and allelopathic activity of the essential oil from *Callistemon viminalis* (Myrtaceae) *Blossoms on Lettuce* (*Lactuca sativa* L.) seedlings. American Journal of Plant Sciences [Internet]. 2014;**5**(24):3551-3557. Available from: http://www.scirp.org/journal/PaperInformation. aspx?PaperID=51896

[46] Mancini E, Arnold NA, de Feo V, Formisano C, Rigano D, Piozzi F, Senatore F. Phytotoxic effects of essential oils of *Nepeta curviflora* Boiss. and *Nepeta nuda* L. subsp. albiflora growing wild in Lebanon. Journal of Plant Interactions. 2009;**4**(4):253-259

[47] Ibrahim MA, Kainulainen P, Aflatuni A, Tiilikkala K, Holopainen JK. Insecticidal, repellent, antimicrobial activity and phytotoxicity of essential oils: With special reference to limonene and its suitability for control of insect pests. Agricultural and Food Science in Finland. 2001;**10**(April):243-259

[48] De Martino L, Mancini E, De Almeida LFR, De Feo V. The antigerminative activity of twenty-seven monoterpenes. Molecules. 2010;**15**(9):6630-6637

[49] Hiltpold I, Turlings TCJ. Belowground chemical signaling in maize: When simplicity rhymes with efficiency. Journal of Chemical Ecology [Internet]. May 29, 2008;**34**(5):628-635. Available from:. DOI: http://link.springer.com/10.1007/s10886-008-9467-6

[50] Kobaisy M, Tellez MR, Dayan FE, Duke SO. Phytotoxicity and volatile constituents from leaves of *Callicarpa japonica* Thunb. Phytochemistry [Internet]. 2002 Sep;**61**(1):37-40 Available from: http://linkinghub.elsevier.com/retrieve/pii/S0031942202002078

[51] Wang R, Peng S, Zeng R, Ding LW, Xu Z. Cloning, expression and wounding induction of p-caryophyllene synthèse gene from *Mikania micrantha* H. B. K. and allelopathie potential of ß-caryophyllene. Allelopathy Journal. 2009;**24**(1):35-44

[52] Verdeguer M, Blázquez MA, Boira H. Phytotoxic effects of Lantana camara, *Eucalyptus camaldulensis* and *Eriocephalus africanus* essential oils in weeds of Mediterranean summer crops. Biochemical Systematics and Ecology [Internet]. Oct 2009;**37**(4):362-369. DOI: http://dx.doi.org/10.1016/j.bse.2009.06.003

[53] Macías FA, Galindo JCG, Molinillo JMG, Cutler HG. Allelopathy: Chemistry and mode of action of allelochemicals [Internet]. In: Macías FA, Galindo JCG, Molinillo JMG, Cutler HG, editors. Allelopathy - Chemsitry and Mode of Action and Allelochemicals. New York, Washington, D.C: CRC Press, Taylor & Francis; 2003. 392 p. Available from: https:// books.google.com.br/books/about/Allelopathy.html?id=B0937BBWXjQC&redir_esc=y

[54] Silva MP, Piazza LA, López D, López Rivilli MJ, Turco MD, Cantero JJ, Tourn MG, Scopel AL. Phytotoxic activity in *Flourensia campestris* and isolation of (–)-hamanasic acid A as its active principle compound. Phytochemistry [Internet]. 2012;**77**:140-148. DOI: http://dx.doi.org/10.1016/j.phytochem.2011.09.020

[55]   Ahluwalia V, Sisodia R, Walia S, Sati OP, Kumar J, Kundu A. Chemical analysis of essential oils of *Eupatorium adenophorum* and their antimicrobial, antioxidant and phytotoxic properties. Journal of Pest Science (2004). 2014;**87**(2):341-349

[56]   Mazzafera P. Efeito alelopático do extrato alcoólico do cravo-da-índia e eugenol. Revista Brasileira de Botânica [Internet]. Jun 2003;**26**(2):231-238 Available from: http://www.scielo.br/scielo.php?script=sci_arttext&pid=S0100-84042003000200011

[57]   Bainard LD, Isman MB, Upadhyaya MK. Phytotoxicity of clove oil and its primary constituent eugenol and the role of leaf epicuticular wax in the susceptibility to these essential oils. Weed Science [Internet]. Oct 20, 2006;**54**(5):833-837 Available from: https://www.cambridge.org/core/product/identifier/S0043174500015599/type/journal_article

[58]   Ajayi OE, Appel AG, Henry Y. Phytotoxicity of some essential oil components to cowpea (*Vigna unguiculata* (L.) Walp.) seeds. International Journal of Plant Biology & Research. 2014;**2**(4):2-9

[59]   Cruz GS, Wanderley-Teixeira V, Oliveira J V., Correia AA, Breda MO, Alves TJ, Cunha FM, Teixeira AA, Dutra KA, Navarro DM. Bioactivity of *Piper hispidinervum* (Piperales: Piperaceae) and *Syzygium aromaticum* (Myrtales: Myrtaceae) oils, with or without formulated Bta on the biology and immunology of *Spodoptera frugiperda* (Lepidoptera: Noctuidae). Journal of Economic Entomology [Internet]. 2014;**107**(1):144-153. Available from: http://www.ncbi.nlm.nih.gov/pubmed/24665696

[60]   Shreeya A, Batish DR, Singh HP. Research paper alleopathic effect of aromatic plants: Role of volatile. Journal of Global Biosciences 2016;**5**(7):4386-4395

[61]   Rolli E, Marieschi M, Maietti S, Sacchetti G, Bruni R. Comparative phytotoxicity of 25 essential oils on pre- and post-emergence development of *Solanum lycopersicum* L.: A multivariate approach. Industrial Crops and Products [Internet]. 2014;**60**:280-90. DOI: http://dx.doi.org/10.1016/j.indcrop.2014.06.021

[62]   Andrés MF, Rossa GE, Cassel E, Vargas RMF, Santana O, Díaz CE, González-Coloma A. Biocidal effects of *Piper hispidinervum* (Piperaceae) essential oil and synergism among its main components. Food and Chemical Toxicology [Internet]. 2017;**109**:1086-1092. DOI: http://dx.doi.org/10.1016/j.fct.2017.04.017

[63]   Meepagala KM, Sturtz G, Wedge DE, Schrader KK, Duke SO. Phytotoxic and antifungal compounds from two Apiaceae species, *Lomatium californicum* and *Ligusticum hultenii*, rich sources of Z-ligustilide and apiol, respectively. Journal of Chemical Ecology. 2005;**31**(7):1567-1578

[64]   Vasilakoglou I, Dhima K, Paschalidis K, Ritzoulis C. Herbicidal potential on *Lolium rigidum* of nineteen major essential oil components and their synergy. Journal of Essential Oil Research. 2013;**25**(1):1-10

Chapter 6

Production and Stabilization of Mycoherbicides

Alexander Berestetskiy and Sofia Sokornova

Additional information is available at the end of the chapter

http://dx.doi.org/10.5772/intechopen.76936

Abstract

Despite the urgent need for alternatives to chemicals in plant protection, biological her-bicides are not widely used as biofungicides and bioinsecticides. The review is devoted to connections between fungal biology, biochemistry, their ability to survive in extreme environment and development of effective mycoherbicides. Advanced studies on the production and stabilization of mycofungicides and mycoinsecticides were analyzed too in order to obtain ideas for the improvement of efficacy and technology of mycoherbi-cides in the future. The analysis of research data published within last 20 years showed following trends. First, more attention is paid for production both effective and stress tolerant propagules especially based on the submerged fungal mycelium and its modi-fications (blastospores, chlamydospores and microsclerotia). Second, the construction of bioreactors, in particular, for solid-state fermentation is continuously being improved that allows producing highly stress tolerant fungal aerial conidia. Third, based on studies of biochemical mechanisms of viability of fungi in extreme environment, the approaches of stabilization and storage of fungal propagules were developed. However, the positive reply to the question, if biopesticides including mycoherbicides, will become a serious alternative to agrochemicals, will be possible when they demonstrate stable efficacy in the field conditions and safety for both environment and end users.

Keywords: biopesticides, fungi, biology, biochemistry, ecology, stress tolerance, mycoherbicides, mycoinsecticides, mycofungicides, production, stabilization, formulation

1. Introduction

With gradual increase of restrictions for use of chemical pesticides, the role of natural regula-tors of pest organisms including weeds and invasive plants will grow up. The development of weed biocontrol is stimulated due to their increasing resistance to chemical herbicides and slow down development of novel herbicidal active components with new mode of action [1]. There are a few mycoherbicides among biopesticides registered in the last years [2].

Despite biocontrol efficacy is generally lower than application of pesticides, biologicals have some advantages over chemicals: (1) biopesticides can be used for resistance management, especially since may have multiple modes of action, which would reduce the chance of resistance occurring in a particular crop pest; (2) many biopesticides have no or low restricted entry intervals, meaning that post-application, restricted entry into the field is very low and there are often no limitations prior to harvest and (3) there are generally exemptions of biopesticides from maximum residue limits because they are considered acceptable and relatively safe [3].

More than half a century had passed since the first mycoherbicide was registered. Dispute raged, and still rages today, about whether "Have bioherbicides come of age?", "What is they really contribution to crop protection?" or "Athletes foot or Achilles heel?" [4–9]. This is partly because the biological herbicides as distinct from chemical preparations are not "stand alone" products. There are significant differences in their origins (biological vs. chemical), modes of action (multiple vs. singular), manufacturing methods (fermentation vs. synthesis), requirements to storing and application conditions, etc. [10]. Efficacy of mycoherbicide strategy depends on thorough understanding of host-pathogen-environment interactions. The biological herbicides are more effective when they are incorporated into integrated weed management programs [11]. For example, it was demonstrated that the bioherbicide *Myrothecium verrucaria* (7.5×10^{12} spores/ha) used along with mowing allows to quickly eradicate kudzu (*Pueraria montana* var. *lobata*) [12]. Bioherbicidal efficacy also can be improved using bio-based formulation [13].

Currently, it highlighted 18 of the most serious weeds in agriculture and 50 troublesome ones in cultivated crops, pastures and waterways [11]. Mycoherbicides are mainly used to prevent and control the spread of such worst parasitic weeds as *Orobanche*, *Phelipanche*, *Striga* and *Cuscuta* [3, 14]. Most of them are invasive species. Invasive plants do not only displace the indigenous species, but also change soil biota over considerable territories. Therefore, the presence of particular arbuscular mycorrhizal fungi may determine the success of their invasion [15–19]. Herbicide contamination also can cause deleterious effects on soil biota. Therefore, it is supposed that mycoherbicides used along with other biological and mechanical methods of plant protection might make a more positive impact on restoring native plants population than chemicals. For example, it was shown the promotion effects of *Fusarium oxysporum* f. sp. *strigae*, a soil-borne biocontrol agent against *Striga hermonthica*, on total fungal and arbuscular mycorrhizal fungal taxa in rhizospheres native plants *Gigaspora margarita* [20, 21]. Application of mycoinsecticide *Metarhizium anisopliae*, for leafroller (*Cnaphalocrocis medinalis*) control increased the relative distribution of bacterial species (*Methylobacterium*, *Sphingobium* and *Deinococcus*) implicated in organic pollutant degradation and plant growth promotion [22].

Key features of mycoherbicides are host specificity, crop tolerance, efficacy, environmental fate, temperature and moisture spectrum, mode of action and toxicology [23]. It is important to realize that not only the choice of the strain, but also types of propagules (conidia, mycelium, sclerotia, etc.), production and application method is influenced by mycoherbicide features. Fungal propagules are influenced by a number of environmental factors (temperature, humidity etc.) that affect their biocontrol efficacy. It was demonstrated that the propagules' choice, formulation and application strategy potentially reduce the dew period requirement [24, 25]. Another possible approach would be a manipulation with fermentation conditions up to product infection materials with set-up parameters [7, 26]. Similarly, during fungal growth physical, chemical and nutritional conditions can be altered to manipulate endogenous reserves for production of propagules with improved stress tolerance to abiotic factors and virulence to host [7, 27–29]. Depending on production method conidia significantly differ by the content of compatible solutes and resistance to environmental influences. The maximum difference is observed when comparing conidia obtained on artificial nutrient media and in nature [28, 30].

Despite of considerable progress in technologies of production and application of mycoherbicides, biopesticides for control of phytophagous insects and plant pathogens have showed much higher commercial success. In some cases, the useful experience for development commercially viable mycoinsecticides and mycofungicides can be tested for the improvement of potential mycoherbicides. For this reason, in this review we analyzed the approaches for producing both mycoherbicides and other types of biopesticides based on fungi.

2. Production

2.1. Choice of propagule types

Various kinds of fungal propagules often fulfill different purposes. In nature, the typical infectious propagules of the pathogenic Ascomycetes are the aerial conidia that facilitate distribution and spreading of these fungi. Generally, aerial conidia can be cost-effectively produced under laboratory conditions [31]. Blastospores, submerged (microcycle) conidia, sporogenically competent mycelia and microsclerotia may be used as the infectious agents as well. They often have a higher survival capability as well as the increased genetic diversity, which probably enhances survival in unstable environments [32, 33]. The morphological and physiological features of submerged conidia can significantly differ from properties of aerial conidia produced by a solid-state culture. For example, submerged conidia and blastospores of *Metarhizium anisopliae* var. *acridum* is characterized by lower surface-hydrophobicity and faster germination as compared to air conidia [34, 35]. Choice of the appropriate propagule is defined by the quality specifications (life-time requirements, desiccation, thermal and UV tolerance, speed of germination and infection, environmental stability and reproduction and the inherent ability) [26, 27, 36–38]. If the conidia production is technologically quite complex or expensive (e.g. due to UV requirements, low viability of propagules during storage and drying, expensive substrates, low-yield spore production, etc.), the mycelium is used as infection material. Application of vegetative mycelium was more effective than conidia in several "fungus-weed" pathosystems *Alternaria cassiae* Jurair & Khan/*Cassia obtusifolia* L. [39], *Chondrostereum purpureum* (Pers. ex Fr.) Pouzar/*Prunus serotina* Erhr. [40], *Phoma herbarum* Westend/*Taraxacum officinale* G.H. Weber ex Wiggers [41], *Sphaceloma poinsettiae* Jenkins & Ruehle/*Euphobia heterophylla* L. [42], *Stagonospora cirsii/Cirsium arvense* [43] and *Alternaria alternata* (Fr.) Keissler/*Eupatorium adenophorum* Spreng. [44]. Possibly, in some cases the fungal mycelium is able to complete the infection process faster than conidia [44, 45]. At the same time, the mycelium is generally less tolerant to the abiotic stress. Nevertheless, the mycelium modifications like chlamydospores, microsclerotia and sclerotia can keep vitality of the fungus for a longer time and can infect the host under suitable weather conditions [46]. Fungal chlamydospores and microsclerotia are evaluated as infection materials for mycoherbicides as well as for other mycopesticides. In nature, chlamydospores formed by *Fusarium oxysporum* play a significant role in long-term survival of the fungus due to their resistance to temperature extremes and desiccation [47, 48]. Chlamydospores of *F. oxysporum* are more thermotolerant than microconidia, it makes them suitable for dry mycoherbicidal formulation. A liquid culture medium was developed for their production [47–49]. A formulation based on dried chlamydospores *F. oxysporum* f sp. *strigae* was developed to control *Striga hermonthica* and *S. asiatica*. It was registered in 2008 in Africa [5, 20]. Another mycoherbicide, DeVine is a liquid formulation of *Phytophthora palmivora* (*P. citrophthora*) chlamydospores for control of milkweed vine (*Morrenia odorata*) in Florida citrus groves. One of the possible weaknesses of such propagules is the uneven germination. Arabic gum in a liquid formulation of chlamydospores of *F. oxysporum* stimulated germ tube elongation and the production of secondary chlamydospores [52]. Nevertheless, the germination rate of conidia of *Rhynchsporium alimatis* was two times lower than the germination of chlamydospores [53]. In the practice, conidia of *Mycoleptodiscus terrestris* cannot be produced using submerged fermentation. At the same time, microsclerotia of this fungus are capable to remain stable in dry conditions and to germinate both hyphally and sporogenically upon rehydration that enhances the potential of this fungus for its use as biological control agent for hydrilla [54]. The mycoherbicide Sarritor was developed on the base of microsclerotia of *Sclerotinia minor*. It demonstrated its high efficacy against dandelion (*Taraxacum officinale*) (78 and 97% by pre- or post-emergence application correspondingly) [55]. The microsclerotia of *Colletotrichum truncatum* can be produced by both submerged and solid-state fermentation and to be effectively used for *Sesbania exaltata* control [56–58]. The development of "multi-propagule" formulations of mycoherbicides is possible as well [58].

However, a few of successful field experiments with microsclerotia-based mycoinsecticide were described. The field efficacy of solid and liquid formulations of microslecrotia *Metarhizium brunneum* F52 was lower or comparable with its conidial preparations. However, microsclerotia of the fungus can be applied with a harsh hydro-mulch technique [59].

2.2. Mass production of mycoherbicide propagules

High spore density (about 10^{12}–10^{14} CFU per ha) is required for use of mycoherbicides in the field. Therefore, one of the main technological goals is to obtain cost-effective, viable and aggressive infectious material [3, 26]. The secondary use of substrates is a solution of their decontamination and utilization. For example, multi-step waste wood bio-recycling includes the cultivation of *Lentinula edodes* and *Pleurotus ostreatus* followed by *Trichoderma, Beauveria* and *Brachycladium* biocontrol strains [60–62].

The loss of viability of the infectious material is usually observed during its drying and storage. Moreover, in nature, the combination of temperature and humidity optimal for rapid germination of fungal spores is relatively rare. Germination of spores can be also suppressed by the action of solar irradiation. Thus, the techniques and conditions for cultivation of biocontrol fungi and the selection of the nutrient media composition should be directed both to reach high biomass yields and to improve their activity in the field [63, 64].

There are several approaches to improve fitness of biocontrol fungi: strain selection, optimization of media composition, addition protectors (compatible solutes such as trehalose, sucrose, glycine-betaine, etc.) and treatment of growing cultures with sub-lethal doses of stress factors (e.g. oxidative stress and temperature) [26, 65, 66]. However, on the practice sub-optimal water activity of the substrates are widely used and helpful [67, 68].

Propagules can be produced by solid-state and liquid fermentation or two-phase system.

2.2.1. Liquid submerged fermentation (LSF)

LSF is the most commonly used technology for microbial inoculum production. Collego and DeVine, the first commercially produced bioherbicides, had been manufactured this way. The ability to fully control the cultivation process and its relatively short duration (several days) is an undoubted advantage of LSF over solid-state fermentation. The composition of a culture medium is an important parameter in the biotechnological process because it is 30–40% of the production costs. A commercial LSF medium for *C. truncatum* conidia production includes glucose (20 g/l), yeast extract (2.5 g/l), cottonseed flour (7.5 g/l) and various salts. After 72 h cultivation, more than 6×10^7 conidia/ml is produced [69]. To obtain a high titer of *Paecilomyces fumosoroseus* blastospores resistant to lyophilization, a nutrient medium was optimized, allowing to receive $1–2 \times 10^9$ spores/ml after a 48-h fermentation. The key factors were high inoculum concentration, amino acid-rich nitrogen source and trace elements [70]. The nutrient medium composition and fermentation parameters (2% of inoculum, duration 120–160 h) for production of mature chlamydospores (more than 1×10^8 CFU/ml) of Gliocladium virens GL-21 (SoilGard™ biofungicide) were selected [71].

To obtain a high yield of viable and stress tolerant infectious material, the composition of the liquid nutrient medium requires optimization. Its algorithm can include three main steps: (1) selection of the basal medium with a set of vitamins and trace elements, on which the fungus grows and/or sporulate well; (2) selection of carbon and nitrogen sources and their optimal concentration and ratio determination and (3) replacement of artificial carbon and nitrogen sources by cheap natural ones [72]. Application of factorial design and response surface methods were successfully used to optimize the growth parameters required for large scale conidia production of potential mycoherbicides based on *Gloecercospora sorghi* and *Septoria polygonorum* [73, 74].

To obtain high titers of *Colletotrichum coccodes* conidia, the optimal carbon concentration in the medium was 20 g/l and C/N ratio of 10:1 [75]. In the case of *C. truncatum*, microcyclic sporulation was induced at carbon concentration in the medium up to 4–16 g/l and C/N ratio in the range 10:1–80:1. At carbon concentration of more than 25 g/l in the submerged culture of this fungus, microsclerotia were formed. The maximum yield of *C. truncatum* conidia was obtained at carbon concentration up to 4–8 g/l and C/N ratio of 30:1, but conidia from media with C/N ratio of 10:1 were more pathogenic and resistant to drying. The conidia obtained in the latter medium contained more proteins, trehalose and polyols, but less glucose and lipids than from C/N ratios in the range 30–80:1 [64, 76, 77]. The influence of carbon concentration and C/N ratio on fungal growth and sporulation is not only species, but also strain dependent [78].

The liquid nutrient medium tonicity has a significant effect on the yield and quality of propagules. Sporulation of *Ulocladium atrum* in a liquid medium was obtained with a reduced water potential (Ψ = −2.1 MPa) by adding glycerol (7.3% w/v) and calcium chloride (20 mM) to the medium. Biomass from liquid cultures responded to water-stress by accumulating increased concentrations of polyols (glycerol) and trehalose [79]. Increased liquid nutrient medium tonicity (osmolality 804–1454 mOsm) of the made by 50–150 g/l of PEG 200 polyethylene glycol increased the yield of submerged *Metarhizium anisopliae* var. *acridum* conidia up to 25%. Spores from high osmolality medium had increased pathogenicity and tolerance to drying. Interestingly, relative drying stability did not appear to be the result of differences in polyol or trehalose concentrations [35].

Non-optimal carbon sources also stimulated *M. anisopliae* formatting resistant to long-range ultraviolet (<290 nm) and accumulating about two times more mannitol and trehalose conidia [80]. The effect of alkane-growth induction of the entomopathogenic fungus *Beauveria bassiana* on the virulence was demonstrated. That alkane supplementation of culture media does not affect the fatty acid composition but change the unsaturated/saturated ratio. However, the unsaturated/saturated ratio diminished markedly from 4.32 to 2.47 [81].

At the same time, liquid substrates are uncommon one for fungal growth.

2.2.2. Solid-state fermentation (SSF)

Solid-state fermentation is the most suitable for cultivation of fungi because their habitats are chiefly solid substrates. In fact, SSF imitates the yields aerial conidia as the final product of conidiation processes. For example, 98% of marine fungi were isolated from submerged solid substrates [82]. In the most cases, spore yields and viability are higher than they are produced by SSF [83]. Hydrophobic air conidia are best suitable for oil formulations, since prolong the conidial viability and decreases UV radiation sensitivity [84–86]. Indeed, numerous studies have shown that conidia produced in an SSF culture are tolerant toward environmental factors (dehydration, drop of temperature and solar irradiation) than spores obtained by SmF [87]. Conidia and blastospores are the main infective units used in biological control with entomopathogenic fungi. There is no absolute advantage between both infective units. However, most formulations of mycoinsecticides are based on aerial conidia obtained in solid-state culture, since these propagules are more resistant to abiotic factors found in open fields [88].

A polysaccharide matrix often surrounds the spores produced by SSF and protects them during desiccation opposite the spores produced by LSF [89]. The choice of substrate, its humidity and growing time also affect the quality of propagules [90]. For example, dried conidia of Colletotrichum truncatum produced on vermiculite tended to retain efficacy during storage better than spores recovered from perlite culture [91]. Sometimes the fermentation can be terminated after the fungus has penetrated the nutritive substrate but before conidiation has begun [92, 93]. Dried grain kernels colonized by *Beauveria* or *Metarhizium* remain competent for regrowth and sporulation upon rehydration. The colonized grains are also viable for lengthy periods in the soil, germinating when suitable conditions arise. For example, after such granules are applied into soil or mixed into horticultural soil the conidia were produced within the habitat of the target insect [92, 93]. At the same time, SSF is not widely used earlier in bioherbicide production due to higher costs, more chances of contamination and the complexity of spores' recovery from the substrate [94].

In the case of small manufacturers, the propagules traditionally produced in the plastic bottle or perforated polypropylene carrier bags [95, 96]. This process was the first designed to meet the biological requirements of genus *Metarhizium* fungi. Technology allowed obtaining the conidial yield 1.5×10^9 conidia/g rice and substrate handling capacity was 82 kg rice/production cycle [97]. Later it has been used to produce conidia of the other entomopathogenic fungi like *Beauveria bassiana, Lecanicillium lecanii* and *Penicillium frequentans*, and phytopathogenic fungus *Lasiodiplodia pseudotheobromae* [98–101]. However, this process presents difficulties in terms of monitoring and various process parameters control, which directly affect the production yields and quality. These problems are already apparent at small scale in the laboratory and are exacerbated with increase in scale. For example, in the most of SSF bioreactors constructions it is extremely difficult to eliminate the temperature gradient and the oxygen concentration in the substrate [83, 95]. Elevating of CO_2 levels in substrate can suppress conidiation of *Alternaria cassiae* and *A. crassa* [102]. These problems can be solved by selecting water-retaining additives to the substrate (e.g. cannabis trusses), appropriate stirring and aeration of the substrate [96]. Use of tray bioreactors results in similar or higher production and productivity of conidia than those obtained with the traditional. That is now possible because of advances in the construction of SSF bioreactors [87, 104]. For example, a stirred bioreactor with aeration supply has been designed for *Paecilomyces lilacinus* conidia production [103]. Laverlam International Corporation developed *B. bassiana* in SSF column bioreactors. As well as traditional approach shows the process consists of biphasic system. Producing by submerge fermentation inoculum is used for SSF [97]. The method was developed to produce conidia *Coniothyrium minitans* in internally agitated bioreactor on the oats, as substrate, providing the volumetric conidia yield more than 5×10^{14} conidia/m^3. Significant yield increase probably is provided for the internal agitation caused mechanical damage to mycelium, which directly affected conidia production [105].

It is well known the positive effect of near ultraviolet radiation on sporulation of certain phytopathogenic fungi from genera *Ascochyta* and *Alternaria*. In the application of UV during fermentation and the employment of microbial mixed cultures, SSF can offer this option that cannot be achieved by SmF. However, a direct comparison between the SSF and SmF cultivation modes of fungi is difficult to make because the two processes differ [82].

Naturally occurring substances can be applied for bioherbicide production [106, 107]. SSF allows to obtain bioherbicides utilizing the agroindustry waste such as bagasse, soybean bran and corn steep liquor [108].

3. Stabilization of fungal propagules

Biological material produced by fermentation and separation from a substrate as a rule cannot be stored for a long time. Even at a low temperature of the storage fungal spores, the mycelium can germinate slowly under appropriate wetness that is unpromising without a plant substrate. Many locally produced biopesticides should be used within several weeks after fermentation was finished as DeVine™, a mycoherbicide based on spores of *Phytophthora palmivora* [64, 109].

At the high-productivity biotech companies, the microorganisms should be stabilized to prevent germination of propagules for a long time (months, years). This can be achieved basically by concentration, drying or encapsulation of biomaterial on polymer layer and storage under appropriate conditions. In the ideal situation, the modern biopesticides can be stored not less than 2 years at the temperature 4°C, 3 months at 30°C and several days at 40–50°C [64].

There are quite simple and cheap techniques of stabilization and storage of some microorganisms. For instance, infection material of *Fusarium oxysporum* antagonistic strains is produced, dried and stored in the peat. The fungal spores did not lose viability for several years [110]. There are no universal recipes. An optimal stabilization technique should be developed for any fungal biocontrol agent.

It is well known that fungal growth and development are depend on temperature, free water availability, pH and oxygen concentration. For stabilization of the fungal propagules, these factors are manipulated by lowering pH, water activity, temperature and oxygen concentration [67, 68].

In many fungi, spores or spore matrix contains the inhibitors that prevent their germination in fruiting bodies, conidiomata, pustules even at the favorable wetness and temperature. These compounds isolated from some rust and anthracnose fungi were demonstrated to be fungistatic [111–115]. Probably, they can be used as natural preservatives and for stabilization of spores of biocontrol fungi.

Spores of many different fungi aggregated in conidiomata can survive over a season and longer under stress and varied environmental conditions including drying, UV-irradiation and low winter temperature. As a rule, such spores are pigmented or/and surrounded by thin shell (as teliospores of rust and smut fungi) or incorporated into spore matrix (as in coelomycetous fungi). Chemical analysis of the matrix in *Ascochyta* and *Phoma* spp. showed that it consists of pigments, glucose, polysaccharides, tyrosine and proteins [116, 117].

Protective compounds, such as pigments and compatible solutes, in fungal cells as well thickness of cell wall and plasma membrane lipid composition play important role in their resistance to artificial drying. Pigments, especially phenolic ones, utilize reactive oxygen species (ROS) which production is induced in drying process [28, 46]. Taking in account this consideration protective compounds are added to the biomaterial (at the concentration about 5–20%) before drying to prevent deleterious effects of ROS and to regulate osmotic pressure. Dried biomass should be stored at the darkness and lower oxygen concentrations. The rehydration is the important step too. It should be gradual and be made in wet atmosphere, warm water (30–37°C) in order to prevent the injury of fungal plasma membranes [46, 118].

3.1. Biomass concentration and preservation

The preparation of the concentrated suspensions or emulsions, pastes with addition of preservatives (germination inhibitors, antibiotics, etc.) is the simple techniques of stabilization and storage of fungal propagules, especially, if the it sensitive to drying.

3.1.1. Concentrates

A liquid formulation of the biofungicide was developed on the base of the yeast *Rhodotorula minuta* for biocontrol of mango anthracnose. The addition of glycerol (20%) and xanthan

(5%) to the concentrated spore suspension (10^9 CFU/mL) prevented a preparation contamination and cell sedimentation. At the temperature 4°C, CFU number was decreased 100 times after 6 months of the storage. For stabilization of a bioinsecticide based on the mycelium of *Lagenidium giganteum* a concentrated emulsion was developed containing 40% of refined corn oil and 0.5% AEROSIL (Fumed Silica, R974). The latter prevented mycelium sedimentation and aggregation. This formulation can be stored under room conditions for 12 weeks without loss of efficacy against mosquitos [119].

Some components of emulsion concentrates (for instance, plant or paraffinic oils) affect efficacy of biopesticides including mycoherbicides. They prevent fast water evaporation from spray droplets and improve thermotolerance of fungal cells as it was shown for *Metarhizium anisopliae* s.l. (IP 46) and *Metarhizium robertsii* (ARSEF2575) [120]. Application of *Microsphaeropsis amaranthi* against the weed, *Microsphaeropsis amaranthi* in Sunspray 6E oil (10% v/v) resulted in improved disease impact under low moisture conditions [121].

3.1.2. Pastes

The mycelium of *Trichoderma asperellum* GSS 03-35 produced by submerged liquid fermentation was stabilized by concentration to 6–10% of dry matter into paste containing corn starch (5%) as stiffener. The paste had pH 3 and contained copper sulfate (20 mg/L). During the course of storage, the fungus formed chlamydospores and conidia. After 6 months of storage at the temperature 20°C the fungus remained effective against head blight of wheat [122].

3.2. Drying

The drying is the most popular technique of inoculum stabilization. Besides simple drying by warm heat on trays (convection drying), spray drying, fluid bed drying and lyophilisation (freeze-drying) are used. The selection of the drying technique depends on availability, costs and sensitivity of the biomaterial.

3.2.1. Convection drying

The biomaterial mixed with preservatives and fillers is dried on trays in thin layer. This technique is used for production of the biofungicide *Trichodermin* (Biotechmash, the Ukraine) as a wettable powder. This simple technique can be applied for drying of low-scale amounts of the biomass and for preliminary experiments. Corn starch, rice flour, talc, diatomaceous earth and kaolin were evaluated as preservatives and fillers during drying of blastospores of *Beauveria bassiana*. Kaolin (at the concentration of 5% w/v of spore concentrate) allowed to maintain satisfactory viability of spores (≥50%) for 7 weeks storage at 4°C [123]. Conidia of *Stagonospora convolvuli* LA39 produced on V8 agar and dried with kaolin as a filler (1 g per 10^9 conidia) by air flow kept high viability (>70%) and pathogenicity about 5 months under the temperature 3°C. After 17 months of the storage just 5% of total conidia were viable, when the conidia were stored under the temperature 20°C conidia their viability decrease to 20% for a week [124].

In some inoculum stabilization protocols, convection drying was proposed for formulation of conidia and microsclerotia of *Beauveria*, *Metarhizium*, *Colletotrichum*, *Mycoleptodiscus* and *Trichoderma*. Some useful additives can be used to improve of viability of the infection units:

skimmed milk or/and glycerol (ca. 1–2%, nutrient sources, humectants), clay (ca. 5%, kaolin or peat to protect conidia against UV) and plant oil (4–10%, adhesive) [126, 127].

The drying technique "Stareze" is based on the addition of a membrane stabilizer (sucrose) during the fermentation. High concentration of sucrose (400 g/L) was added to 96-h submerged culture of *Metarhizium anisopliae*. The fermentation was stopped after 168 h and a filler (silica, HiSil™-233, 35 g/L) was added. The filtered product was dried on the trays at ambient temperature. The blastospores of *M. anisopliae* stayed alive for 6 months at 2–4°C [128].

3.2.2. Spray drying

The spore suspension with some adjuvants and additives is sprayed in heated air followed by fast drying (5–30 s). In the case of fluid bed drying, the suspension follows to the bed from dried material babbling by air that forms pseudo-boiling layer. Particles of the drying material stick to gradually form granules (www.niroinc.com).

Submerged conidia of *M. anisopliae* mixed with defatted milk (20%) and sucrose (2.5%) survived better spray drying than freeze-drying process. However, inlet and outlet temperature caused significant effect on their viability [129]. Granules of the commercial biofungicide Contans® are produced by drying conidia of *Coniothyrium minitans* in glucose solution in a spray drier; the product contains about 95% of glucose and 5% of conidia (ca. 1×10^{13} conidia/kg) and remains effective for 2 years when stored at 4°C [96]. In some cases, this technique of drying is not appropriate. Conidia of the epiphytic fungus, *Epicoccum nigrum*, produced by solid-state fermentation lost viability after spray drying at inlet temperature 150°C. However, fluid bed drying was favorable: dried at 30–40°C conidia remained viable even without any preservative and can be stored for 90 days and more [130].

A method was developed for microencapsulation of *Trichoderma* conidia with sugar through spray drying. Microencapsulation with sugars, such as sucrose, molasses or glycerol, significantly ($P < 0.05$) increased the survival percentages of conidia after drying. Microencapsulation of conidia with 2% sucrose solution resulted in the highest survival percentage when compared with other sucrose concentrations and had about 7.5×10^{10} CFU in each gram of dried conidia, and 3.4 mg of sucrose added to each gram of dried conidia. The optimal inlet/outlet temperature setting was 60/31°C for spray drying and microencapsulation. The particle size of microencapsulated conidia balls ranged from 10 to 25 μm. The spray dried biomass of *T. harzianum* was a flow-able powder with over 99% conidia, which could be used in a variety of formulation developments from seed coatings to sprayable formulations [131].

3.2.3. Freeze-drying

Under liophylisation, water vapors from ice under low pressure bypass the liquid state. Conidia of *Septoria passiflorae* survived well after freeze-drying when 10% of skimmed milk was added to the conidial suspension. The fungus stayed viable for >1 year when stored in a vacuum package at 1°C [132]. Blastospores of *Paecilomyces fumosoroseus* together with protectors (10% lactose + 1% bovine albumin, or composition of starch, vegetable oil, sucrose and milk) remain viable after freeze-drying at the level 75% for 50 weeks at −20°C, while at 4°C their viability decreased to the level of 10% [133].

3.3. Encapsulation

Concentrated biomaterial can be incorporated into different polymer matrices that protect fungal cells from effects of some factors such as UV-irradiation and microbial contamination. Products that are resulted from encapsulation process include gel, granules, capsules and microcapsules. There are various industrial equipment for their production [134].

3.3.1. Alginate granules

The process is based on the polymerization of sodium alginate in the solution calcium chloride. For instance, the suspension of the biomaterial (1 part) is mixed with sodium alginate (1.3% solution, 4 parts) and kaolin (5% of total weight); the mixture is dropped into 0.25 M solution of calcium chloride; the resulted granules are filtered and dried. The technique was used for the first time to formulate conidia of *Alternaria cassiae* [135]. Intensity of the fungal sporulation on the granules depended on inoculum production method, fillers and adjuvants; kaolin can be effectively replaced by corn flour [136].

Various compositions of alginate granules were evaluated for many potential and commercial biopesticides. Chitin (2% of granules weight) together with wheat bran (2%) significantly increased spore production of *Beauveria bassiana* on the granules [137]. Starch addition accelerated the rupture of granules and colonization of the peat substrate by *Trichoderma* sp. [138]. For field experiments of biofungicide based on the non-toxigenic *Aspergillus flavus* different adjuvants (1% of granules weight) and fungicides (0.5–1.25 mg per 50 g of the mixture of sodium alginate and 2.5 g of corn flour) were evaluated. Triptone and peptone addition significantly stimulated spore production of the fungus on the granules. The fungicides did not inhibited the antagonist development and preserve the granules against contamination [136].

Composition of alginate formulation of *Trichoderma* sp. conidia optimized by factorial design experiments included glycerol (2% w/v), sodium polyphosphate 2% (w/v) and citrus pectin that allows to maintain the satisfactory titer of conidia for 14 weeks at 28°C. Formulation quality was monitored by Fourier-transform infrared spectroscopy and some chemical interactions between polymers were found [138]. For production of complex mycoinsecticides ("attract and kill") based on *Saccharomyces cerevisiae* (used as an attractant for wireworms) and *Metarhizium brunneum* (as an insect killer), the technical scale technology was developed that included jet cutting of droplets and bed drying of granules at 40–50°C till aw 0.1–0.2 [139].

3.3.2. Microencapsulation

Fungal biomaterial (e.g. conidia and mycelia) suspended in sodium alginate solution or in the mixture of agar-agar (1%) and gelatin (1:1, v/v) is emulsified in corn oil with n-hexadecan (6:4) and lecithin as emulsifier. Gelatin-agar globules were gelated in the emulsion while alginate microcapsules were polymerized when dropped into calcium chloride solution. The size of microcapsules varied from 10 to 400 μm depending on ratio of the mentioned components. The microcapsules were separated from the liquids by vacuum filtration and used by spraying. The microencapsulation technique was successfully used in model experiments for development of artificial conidia based on conidia of *Fusarium avenaceum* and mycelium of *Bipolaris sorokiniana* [10–141].

3.3.3. "Pesta" granules

The production of Pesta granules is based on the technology of pasta production. Inoculum suspension (52 mL), wheat semolina flour (80 g) and kaolin (20 g) are mixed to produce dough. The dough is passed through a pasta maker after that it is dried, crashed and sieved. The technique was tried for encapsulation of conidia of potential mycoherbicides (*Alternaria cassiae*, *A. crassa*, *Colletotrichum truncatum* and *Fusarium lateritium*) as well for stabilization of entomopathogenic nematodes [142, 143]. Melanized fungal structures as pigmented conidia, chlamydospores, microsclerotia and sclerotia granules are generally compatible to Pesta process while non-pigmented conidia of *F. oxysporum*, *C. truncatum*, *Trematophoma lignicola* were not viable in the final product [49, 57, 143–145].

Microslecrotia of *C. truncatum* survived in Pesta granules and remained to produce virulent conidia (for biocontrol of weed *Sesbania exaltata*) for 52 weeks at 25°C low water activity (aw 0.18–0.29), and for 10 years at 4°C [57, 146] while the fungal conidia can be stored no more than 32 weeks [147]. Interestingly, that during the process of encapsulation of *Alternaria alternata* conidia with Pesta process, the number of colony forming units increased due to destroying their aggregations. The virulence of the fungus was stable at a low relative air humidity (12%) for more than 2 years [145].

The composition of Pesta granules can be easily modified. Shabana et al. [148] evaluated various compositions for *Fusarium oxysporum* f. sp. *orthoceras* using 3% (w/w) sucrose, corn flour, glycerol, starch WaterLock B209, cellulose and yeast extract. The last component improved viability of chlamydospores as well as of microconidia in the granules. However, the prepared samples showed appropriate viability (60–80% for 12 months) under 25°C and relative humidity 11–12%; under higher temperature (25°C) and humidity (51–53%) viability of the fungus dramatically decreased by the 4–8th month of storage [148]. The biocontrol efficacy of *Aspergillus alliaceus* against parasitic weeds (*Orobanche* spp.) incorporated in Pesta granules was improved by addition of potato broth or sorghum meal [149].

For encapsulation of conidia of potential mycopesticides (*C. truncatum*, *Alternaria* sp., *Paecilomyces fumoroseus*, *Aspergillius flavus*, *A. parasiticus*) produced by solid and liquid state fermentation the twin-screw extrusion was successfully tested. Ingredients were mixed in the mixer of an extruder and resulted Pesta granules were dried by fluid bed drying at 50°C. The inoculum produced by solid-state fermentation was shown to be less sensitive to whole the stabilization process than the biomaterial from the liquid culture [150].

3.3.4. Stabilize granules

The main components of these granules are a membrane stabilizer (for instance, sucrose at the concentration 10–65% from granules weight), a water absorbance agent (starch), a filler (diathomaceous earth, silica Hi-Sil® at the concentration 5–20%). Additionally, the granules can include vegetable oil (ca. 20%), UV-protectant, preservatives and other inert fillers [151]. For example, sucrose (4 parts), starch (1 part), unrefined vegetable oil (1 part), silica gel (1.5 parts) and biological suspension (4 parts) are mixed and extruded; the resulted pasta is conventionally dried and crashed or milled. This technique was successfully used for potential bioherbicides based on *Fusarium oxysporum* (microconidia and mycelium) and *Pseudomonas* spp. that remain viable for a long time [152–154]. However, submerged conidia of *Metarhizium anisopliae* (the producer of the bioinsecticide Green Muscle™) survived better when the above-mentioned process Satreze was used [128].

The safety and evaluation of postponed risks of mycopesticides are still under question. An agroecosystem is inundated by a fungus at very high concentrations and there is a risk of the crop injury. Some plant pathogens can survive in the soil or plant debris. They are able of producing biologically active compounds (mycotoxins, antibiotics, phytotoxins, etc.). The number of safety research on the safety of mycoherbicides is limited to *Sclerotinia sclerotiorum*, *S. minor*, *Colletotrichum coccodes*, *Fusarium oxysporum* f. sp. *strigae*, *Phoma macrostoma* and *Stagonospora convolvuli* [7, 23, 55, 155–159]. The experience of field observations is limited to several years.

Molecular marking of biocontrol strains is an approach for their post-application tracking and quantification. For instance, the strain *Fusarium oxysporum* f. sp. *strigae* F2, which is potential mycoherbicide against *Striga* spp., was compared with several strains *F. oxysporum* using fluorescent AFLP. Based on this comparison a specific PCR primer was developed for making F2 only in the soil [20, 21, 160].

In conclusion, the approaches for stabilization and storage of biopesticides based on fungal propagules were discussed in this review. In order to produce both virulent and stress tolerant propagules for mycoherbicides based on the submerged fungal mycelium as well as on conidia, chlamydospores and microsclerotia a liquid medium should be optimized. The construction of bioreactors, in particular, for solid-state fermentation is continuously being improved that allows of producing highly stress tolerant fungal aerial conidia. Various recipes for liquid (e.g. suspension and emulsion concentrates) and solid (like alginate and stabilize granules) formulation of mycoherbicides were developed to be stored for a long time and effectively used. However, the efficacy of mycoherbicides is still unstable and their safety is not proved clearly to be widely commercialized.

Acknowledgements

The research was supported by Russian Science Foundation (project # 16-16-00085).

Author details

Alexander Berestetskiy* and Sofia Sokornova

*Address all correspondence to: aberestetskiy@vizr.spb.ru

All-Russian Institute of Plant Protection, Saint-Petersburg, Russia

References

[1] Westwood JH, Charudattan R, Duke SO, Fennimore SA, Marrone P, Slaughter DC, Swanton C, Zollinger R. Weed management in 2050: Perspectives on the future of weed science. Weed Science. 2018;**66**(3):275-285. DOI: 10.1017/wsc.2017.78

[2] Van Lenteren JC, Bolckmans K, Köhl J, Ravensberg WJ, Urbaneja A. Biological control using invertebrates and microorganisms: Plenty of new opportunities. BioControl. 2018;**63**:39-59. DOI: 10.1007/s1052

[3] Bailey KL. Canadian innovations in microbial biopesticides. Canadian Journal of Plant Pathology. 2010;**32**(2):113-121. DOI: 10.1080/07060661.2010.484195

[4] Lazarovits G, Goettel M, Vincent C. Adventures in biocontrol. In: Vincent C, Goettel M, Lazarovits G, editors. Biological Control: A Global Perspective. Case Histories from Around the World. Wallingford: CABI Publishing; 2007. pp. 1-6. DOI: 10.1079/9781845932657.0001

[5] Watson A, Gressel G, Sands D, Hallett S, Vurro M, Beed F. *Fusarium oxysporum* f.sp. *Striga*, athletes foot or Achilles heel? In: Vurro M, Gressel J, editors. Novel Biotechnologies for Biocontrol agent Enhancement and Management. Netherlands: Springer; 2007. pp. 1-11. DOI: 10.1007/978-1-4020-5799-1_11

[6] Ash GJ. The science, art and business of successful bioherbicides. Biological Control. 2010;**52**:230-240. DOI: 10.1016/j.biocontrol.2009.08.007

[7] Bailey KL, Falk S. Turning research on microbial bioherbicides into commercial products—A *Phoma* story. Pest Technology. 2011;**5**(Special Issue 1):73-79 http://www.globalsciencebooks.info/Online/GSBOnline/images/2011/PT_5(SI1)/PT_5(SI1)73-79o.pdf

[8] Müller E, Nentwig W. Plant pathogens as biocontrol agents of *Cirsium arvense*—An overestimated approach? NeoBiota. 2011;**11**:1-24. DOI: 10.3897/neobiota.11.1803

[9] Glare T, Caradus J, Gelernter W, Jackson T, Keyhani N, Köhl J, Marrone P, Morin L, Stewart A. Have biopesticides come of age? Trends in Biotechnology. 2012;**30**(5):250-258. DOI: 10.1016/j.tibtech.2012.01.003

[10] Bailey K. Microbial weed control: An off-beat application of plant pathology. Canadian Journal of Plant Pathology. 2004;**26**(3):239-244. DOI: 10.1080/07060660409507140

[11] Bailey KL. The bioherbicide approach to weed control using plant pathogens. In: Abrol DP, editor. Integrated Pest Management: Current Concepts and Ecological Perspective. Vol. 13. San Diego: Elsevier (Academic Press); 2014. pp. 245-266. DOI: 10.1016/B978-0-12-398529-3.00014-2

[12] Weaver MA, Boyette CD, Hoagland RE. Rapid kudzu eradication and switchgrass establishment through herbicide, bioherbicide and integrated programmes. Biocontrol Science and Technology. 2016;**26**(5):640-650. DOI: 10.1080/09583157.2016.1141175

[13] Boyette CD, Hoagland RE, Stetina KC. Efficacy improvement of a bioherbicidal fungus using a formulation-based approach. American Journal of Plant Sciences. 2016;**7**(16):2349-2358. DOI: 10.4236/ajps.2016.716206

[14] Hershenhorn J, Casella F, Vurro M. Weed biocontrol with fungi: Past, present and future. Biocontrol Science and Technology. 2016;**26**(10):1313-1328. DOI: 10.1080/09583157.2016. 1209161

[15] Li HN, Xiao B, Liu WX, Wan FH. Changes in soil biota resulting from growth of the invasive weed, *Ambrosia artemisiifolia* L. (*Compositae*), enhance its success and reduce growth of co-occurring plants. Journal of Integrative Agriculture. 2014;**13**(9):1962-1971. DOI: 10.1016/S2095-3119(13)60569-9

[16] Glushakova AM, Kachalkin AV, Chernov IY. Specific features of the dynamics of epiphytic and soil yeast communities in the thickets of *Indian balsam* on mucky gley soil. Eurasian Soil Science. 2011;**44**(80):886-892. DOI: 10.1134/S1064229311080059

[17] Glushakova AM, Kachalkin AV, Chernov IY. Soil yeast communities under the aggressive invasion of Sosnowskys hogweed (*Heracleum Sosnowskyi*). Eurasian Soil Science. 2015;**48**(2):201-207. DOI: 10.1134/S1064229315020040

[18] Glushakova AM, Kachalkin AV, Chernov IY. The influence of *Aster × salignus* Willd. Invasion on the diversity of soil yeast communities. Eurasian Soil Science. 2016;**49**(7):792-795. DOI: 10.1134/S1064229316050057

[19] Majewska ML, Rola K, Zubek S. The growth and phosphorus acquisition of invasive plants *Rudbeckia laciniata* and *Solidago gigantea* are enhanced by arbuscular mycorrhizal fungi. Mycorrhiza. 2017;**27**(2):83-94. DOI: 10.1007/s00572-016-0729-9

[20] Zimmermann J, Musyoki MK, Cadisch G, Rasche F. Biocontrol agent *Fusarium oxysporum* f.sp. *strigae* has no adverse effect on indigenous total fungal communities and specific AMF taxa in contrasting maize rhizospheres. Fungal Ecology. 2016;**23**:1-10. DOI: 10.1016/j.funeco.2016.05.007

[21] Zimmermann J, Musyoki MK, Cadisch G, Rasche F. Proliferation of the biocontrol agent *Fusarium oxysporum* f. sp. *strigae* and its impact on indigenous rhizosphere fungal communities in maize under different agro-ecologies. Rhizosphere. 2016;**1**:17-25. DOI: 10.1016/j.rhisph.2016.06.002

[22] Hong M, Peng G, Keyhani NO, Xia Y. Application of the entomogenous fungus, *Metarhizium anisopliae*, for leafroller (*Cnaphalocrocis medinalis*) control and its effect on rice phyllosphere microbial diversity. Applied Microbiology and Biotechnology. 2017;**101**(17): 6793-6807. DOI: 10.1007/s00253-017-8390-6

[23] Bailey KL, Falk S, Derby JA, Melzer M, Boland GJ. The effect of fertilizers on the efficacy of the bioherbicide, *Phoma macrostoma*, to control dandelions in turfgrass. Biological Control. 2013;**65**(1):147-151. DOI: 10.1016/j.biocontrol.2013.01.003

[24] Jaronski ST. Ecological factors in the inundative use of fungal entomopathogens. BioControl. 2010;**55**:159-185. DOI: 10.1007/s10526-009-9248-3

[25] Bailey K, Derby J-A, Bourdôt G, Skipp B, Cripps M, Hurrell G, Saville D, Noble A. *Plectosphaerella cucumerina* as a bioherbicide for *Cirsium arvense*: Proof of concept. BioControl. 2017;**62**:693-704. DOI: 10.1007/s10526-017-9819-7

[26] Jaronski ST, Mascarin GM. Mass production of fungal entomopathogens. In: Lacey L, editor. Microbial Control of Insect and Mite Pests. Elsevier (Academic Press). Amsterdam and Boston. Vol. 9. 2017. pp. 141-155. DOI: 10.1016/B978-0-12-803527-6.00009-3

[27] Jackson MA, McGuire MR, Lacey LA, Wraight SP. Liquid culture production of desiccation tolerant blastospores of the bioinsecticidal fungus *Paecilomyces fumosoroseus*. Mycological Research. 1997;**101**:35-41. DOI: 10.1017/S0953756296002067

[28] Magan N. Physiological approaches to improving ecological fitness of fungal biocontrol agents. In: Butt TM, Jackson CW, Magan N, editors. Fungi as Biocontrol Agents: Progress, Problems and Potential. Wallingford: CABI Publishing; 2001. pp. 239-252. DOI: 10.1079%2F9780851993560.0239

[29] Rangel DE, Braga GU, Fernandes ÉK, Keyser CA, Hallsworth JE, Roberts DW. Stress tolerance and virulence of insect-pathogenic fungi are determined by environmental conditions during conidial formation. Current Genetics. 2015;**61**(3):383-404. DOI: 10.1007/s00294-015-0477-y

[30] Shah FA, Wang CS, Butt TM. Nutrition influences growth and virulence of the insect-pathogenic fungus *Metarhizium anisopliae*. FEMS Microbiology Letters. 2005;**251**(2):259-266. DOI: 10.1016/j.femsle.2005.08.010

[31] Templeton GE, TeBeest DO, Smith JRJ. Biological weed control with mycoherbicides. Annual Review of Phytopathology. 1979;**17**(1):301-310. DOI: 10.1146/annurev.py.17.090179.001505

[32] Chitarra GS, Dijksterhuis J. The germinating spore as a contaminating vehicle. In: Dijksterhuis J, Samson RA, editors. Food Mycology. A Multifaceted Approach to Fungi and Food. Boca Raton: CRC/Taylor & Francis; 2007. pp. 83-100

[33] Hoekstra RF. Evolutionary biology: Why sex is good. Nature. 2005;**434**(7033):571-573. DOI: 10.1038/434571a

[34] Leland JE, Mullins DE, Vaughan LJ, Warren HL. Effects of media composition on submerged culture spores of the entomopathogenic fungus, *Metarhizium anisopliae* var. *acridum*, part 1: Comparison of cell wall characteristics and drying stability among three spore types. Biocontrol Science and Technology. 2005;**15**(4):379-392. DOI: 10.1080/09583150400016928

[35] Leland JE, Mullins DE, Vaughan LJ, Warren HL. Effects of media composition on submerged culture spores of the entomopathogenic fungus, *Metarhizium anisopliae* var. *acridum*, Part 2: Effects of media osmolality on cell wall characteristics, carbohydrate concentrations, drying stability, and pathogenicity. Biocontrol Science and Technology. 2005;**15**(4):393-409. DOI: 10.1080/09583150400016910

[36] Jackson MA, Dunlap CA, Jaronski ST. Ecological considerations in producing and formulating fungal entomopathogens for use in insect biocontrol. BioControl. 2010;**55**:129-145. DOI: 10.1007/s10526-009-9240-y

[37] Vega FE, Jackson MA, McGuire MR. Germination of conidia and blastospores of *Paecilomyces fumosoroseus* on the cuticle of the silverleaf whitefly, *Bemisia argentifolii*. Mycopathologia. 1999;**147**:33-35. DOI: 10.1023/A:1007011801491

[38] Fernandes ÉKK, Rangel DEN, Braga GUL, Roberts DW. Tolerance of entomopathogenic fungi to ultraviolet radiation: A review on screening of strains and their formulation. Current Genetics. 2015;**61**:427-440. DOI: 10.1007/s00294-015-0492-z

[39] Stowell LJ, Nette K, Heath B, Shutter R. Fermentation alternatives for commercial production of a mycoherbicide. In: Demain AL, Somkuti GA, Hunter-Cevera JC, Rossmoore HW, editors. Novel Microbial Products for Medicine and Agriculture. Society for Industrial Microbiology. Amsterdam: Elsevier; 1989. pp. 219-227

[40] Scheepens PC, Hoogerbrugge A. Control of *Prunus serotina* in forests with the endemic fungus *Chondrostereum purpureum*. In: Proc. VIIth International Symposium Biological Control Weeds; 5-11 March 1988; Roma, Italy, Delfosse, I.S. (ed.). Int. Sper. Pathol. Veg. (MAF). 1989. pp. 545-551. <http://www.bfs.wur.nl/NR/rdonlyres/E158C176-C828-44D7-B278-89B36EE4CAA9/52147/Scheepens.pdf>

[41] Stewart-Wade SM, Boland GJ. Selected cultural and environmental parameters influence disease severity of dandelion caused by the potential bioherbicide fungi, *Phoma herbarum* and *Phoma exigua*. Biocontrol Science and Technology. 2004;**14**:561-569. DOI: 10.1080/09583150410001682296

[42] De Lima Nechet K, Barreto RW, Mizubuti ESG. *Sphaceloma poinsettiae* as a potential biological control agent for wild poinsettia (*Euphobia heterophylla*). Biological Control. 2004;**30**:556-565. DOI: 10.1016/j.biocontrol.2004.03.007

[43] Berestetskiy AO, Kungurtseva OV, Sokornova SV. Can mycelial inoculum be an alternative to conidia in the case of *Stagonospora cirsii* J.J. Davis, a potential biocontrol agent of *Cirsium arvense*? In: Current Status and Future Prospects in Bioherbicide Research and Product Development; 19 June. Vol. 2005. Bari, Italy: Joint Workshop International Bioherbicide Group and EWRS-Biocontrol Working Group; 2005. p. 7

[44] Qiang S, Zhu Y, Summerell BA, Li Y. Mycelium of *Alternaria alternata* as a potential biological control agent for *Eupatorium adenophorum*. Biocontrol Science and Technology. 2006;**16**(7):653-668. DOI: 10.1080/09583150600699804

[45] Sokornova SV, Hutty AV, Berestetskiy AO. The process of infection of the tubercle field with conidia and mycelium of the phytopathogenic fungus *Stagonospora cirsii*. Plant Protection News. 2011;**3**:57-60 (In Russian)

[46] Hoffman P. Factors influencing survival of dried organisms. In: Koch E, Leinonen P, editors. Formulation of microbial inoculants, 5-6 February, 2001. Proc. of COST Action 830 Meeting. Germany: Braunschweig; 2011. pp. 12-15

[47] Nash SM, Christou T, Snyder WC. Existence of *Fusarium solani* f. *phaseoli* as chlamydospores in soil. Phytopathology. 1961;**51**:308-312

[48] Schippers B, Van Eck WH. Formation and survival of chlamydospores in *Fusarium*. In: Nelson PE, Tousson TA, Cook RJ, editors. Fusarium, Diseases, Biology and Taxonomy. University Park, USA: The Pennsylvania State University Press; 1981. pp. 250-260

[49] Müller-Stöver D, Kroschel J, Thomas H, Sauerborn J. Chlamydospores of *Fusarium oxysporum* Schlecht. f.sp. *orthoceras* (APel & Wollenw.) Bilai as inoculum for wheat flour–kaolin

granules to be used for the biological control of *Orobanche cumana* Wallr. European Journal of Plant Pathology. 2002;**108**:221-228

[50] Müller-Stöver D, Thomas H, Sauerborn J, Kroschel J. Two granular formulations of *Fusarium oxysporum* f.sp. *orthoceras* to mitigate sunflower broomrape *Orobanche cumana*. BioControl. 2004;**49**(5):595-602. DOI: 10.1023/B:BICO.0000036438.66150.21

[51] Elzein A, Kroschel J. Influence of agricultural by-products in liquid culture on chlamydospore production by the potential mycoherbicide *Fusarium oxysporum* Foxy 2. Biocontrol Science and Technology. 2004;**14**(8):823-836. DOI: 10.1023/A:1015104903622

[52] Ciotola M, DiTommaso A, Watson AK. Chlamydospore production, inoculation methods and pathogenicity of *Fusarium oxysporum* M12-4A, a biocontrol for *Striga hermonthica*. Biocontrol Science and Technology. 2000;**10**(2):129-145. DOI: 10.1080/09583150029279

[53] Cliquet S, Ash G, Cother E. Production of chlamydospores and conidia in submerged culture by *Rhynchosporium alismatis*, a mycoherbicide of *Alismataceae* in rice crops. Biocontrol Science and Technology. 2004;**14**(8):801-810. DOI: 10.1080/09583150410001720671

[54] Shearer JF, Jackson MA. Liquid culturing of microsclerotia of *Mycoleptodiscus terrestris*, a potential biological control agent for the management of hydrilla. Biological Control. 2006;**38**:298-306. DOI: 10.1016/j.biocontrol.2006.04.012

[55] Abu-dieyeh M, Watson A. Efficacy of *Sclerotinia minor* for dandelion control: Effect of dandelion accession, age and grass competition. Weed Research. 2007;**47**(1):63-72. DOI: 10.1111/j.1365-3180.2007.00542.x

[56] Jackson MA, Schisler DA. Liquid culture production of microsclerotia of *Colletotrichum truncatum* for use as bioherbicidal propagules. Mycological Research. 1995;**99**(7):879-884. DOI: 10.1016/S0953-7562(09)80745-4

[57] Connick WJ, Jackson MA, Williams KS, Boyette CD. Stability of microsclerotial inoculum of *Colletotrichum truncatum* encapsulated in wheat flour-kaolin granules. World Journal of Microbiology and Biotechnology. 1997;**13**:549-554. DOI: 10.1023/A:1018517409756

[58] Boyette CD, Abbas HK, Johnson B, Hoagland RE, Weaver MA. Biological control of the weed *Sesbania exaltata* using a microsclerotia formulation of the bioherbicide *Colletotrichum truncatum*. American Journal of Plant Sciences. 2014;**5**:2672-2685. Published Online August 2014 in SciRes. http://www.scirp.org/journal/ajps DOI: 10.4236/ajps.2014.518282

[59] Behle RW, Richmond DS, Jackson MA, Dunlap CA. Evaluation of *Metarhizium brunneum* F52 (Hypocreales: Clavicipitaceae) for control of Japanese beetle larvae in turfgrass. Journal of Economic Entomology. 2015;**108**(4):1587-1595. DOI: 10.1093/jee/tov176

[60] Titova YuA, Khlopunova LB, Korshunov DV. Two-step waste bioconversion by *Pleurotus ostreatus* and *Trichoderma harzianum*. Mikologiya i Fitopatologiya 2002;**36**(5):64-70 (in Rus.)

[61] Titova YuA, Novikova II, Khlopunova LB, Korshunov DV. Trichodermin as a two-step waste bioconversion product and its efficacy against cucumber diseases. Mikologiya i Fitopatologiya. 2002;**36**(4):76-80 (in Rus.)

[62] Novikova II, Titova JA, Krasnobayeva IL, Ryzhankova AV, Titov VS, Semenovich AS. Peculiarities of the strain *Dendryphion penicillatum* 1.39 development on various nutrient substrata. Mikologiya i Fitopatologiya. 2010;**44**(1):71-87 (in Rus.)

[63] Muniz-Paredes F, Miranda-Hernandez F, Loera O. Production of conidia by entomopathogenic fungi: From inoculants to final quality tests. World Journal of Microbiology & Biotechnology. 2017;**33**(3):57. DOI: 10.1007/s11274-017-2229-2

[64] Wraight SP, Jackson MA, de Kock SL. Production, stabilization and formulation of fungal biocontrol agents. In: Butt TM, Jackson C, Magan N, editors. Fungi as Biocontrol Agents. Wallingford, United Kingdom: CAB International; 2001. pp. 253-287. DOI: 10.1079/9780851993560.0253

[65] Daryaei A, Jones EE, Glare TR, Falloon RE. Biological fitness of *Trichoderma* atroviride during long-term storage, after production in different culture conditions. Biocontrol Science and Technology. 2016;**26**(1):86-103. DOI: 10.1080/09583157.2015.1077929

[66] Miranda-Hernández F, Garza-López PM, Loera O. Cellular signaling in cross protection: An alternative to improve mycopesticides. Biological Control. 2016;**103**:196-203. DOI: 10.1016/j.biocontrol.2016.09.007

[67] Magan N. Fungi in extreme environments. In: Kubicek CP, Druzhinina IS, editors. Environmental and Microbial Relationships, The Mycota. Vol. 4. Berlin, Heidelberg: Springer; 2007. pp. 85-103. DOI: 10.1007/978-3-540-71840-6_6

[68] Teshler MP, Ash GJ, Zolotarov Y, Watson AK. Increased shelf life of a bioherbicide through combining modified atmosphere packaging and low temperatures. Biocontrol Science and Technology. 2007;**17**(4):387-400. DOI: 10.1080/09583150701213695

[69] Silman RW, Nelsen TC. Optimization of liquid culture medium for commercial production of *Colletotrichum truncatum*. FEMS Microbiology Letters. 1993;**107**:273-278

[70] Jackson MA, Cliquet S, Iten LB. Media and fermentation processes for the rapid production of high concentrations of stable blastospores of the bioinsecticidal fungus *Paecilomyces fumosoroseus*. Biocontrol Science and Technology. 2003;**13**(1):23-33. DOI: 10.1080/0958315021000054368

[71] Eyal J, Baker CP, Reeder JD, Devane WE, Lumsden RD. Large-scale production of chlamydospores of *Gliocladium virens* strain GL-21 in submerged culture. Journal of Industrial Microbiology and Biotechnology. 1997;**19**:163-168. DOI: 10.1038/sj.jim.2900430

[72] Jackson MA. Optimizing nutritional conditions for the liquid culture production of effective fungal biological control agents. Journal of Industrial Microbiology and Biotechnology. 1997;**19**:180-187. DOI: 10.1038/sj.jim.2900426

[73] Mitchell JK. Development of a submerged-liquid sporulation medium for the potential smartweed bioherbicide *Septoria polygonorum*. Biological Control. 2003;**27**:293-299. DOI: 10.1016/S1049-9644(03)00024-0

[74] Mitchell JK, Njalamimba-Bertsch M, Bradford NR, Birdsong JA. Development of a submerged-luiquid sporulation medium for the johnsongrass bioherbicide *Gloecercospora*

sorghi. Journal of Industrial Microbiology and Biotechnology. 2003;**30**:599-605. DOI: 10.1007/s10295-003-0088-3

[75]   Yu X, Hallet SG, Shepard J, Watson A. Effects of carbon concentration and carbon-to-nitrogen ratio on growth, conidiation, spore germination and efficacy of the potential bio-herbicide *Colletotrichum coccoides.* Journal of Industrial Microbiology and Biotechnology. 1998;**20**:333-338. DOI: 10.1038/sj.jim.2900534

[76]   Montazeri M, Greaves MP. Effects of nutrition on dessication tolerance and virulence of *Colletotrichum truncatum* and *Alternaria alternata* conidia. Biocontrol Science and Technology. 2002;**12**:173-181. DOI: 10.1080/09583150120124432

[77]   Montazeri M, Greaves MP, Magan N. Microscopic and cytochemical analysis of extracellular matrices and endogenous reserves of conidia of *Colletotrichum truncatum* harvested from carbon- and nitrogen-limited cultures. Biocontrol Science and Technology. 2003;**13**(7):643-653. DOI: 10.1080/09583150310001606246

[78]   Gao L, Sun MH, Liu XZ, Che YS. Effects of carbon concentration and carbon to nitrogen ratio on the growth and sporulation of several biocontrol fungi. Mycological Research. 2007;**111**(1):87-92. DOI: 10.1016/j.mycres.2006.07.019

[79]   Frey S, Magan N. Production of the fungal biocontrol agent *Ulocladium atrum* by submerged fermentation: Accumulation of endogenous reserves and shelf-life studies. Applied Microbiology and Biotechnology. 2001;**56**:372-377. DOI: 10.1007/s002530100657

[80]   Rangel DEN, Anderson AJ, Roberts DW. Growth of *Metarhizium anisopliae* on non-preferred carbon sources yields conidia with increased UV-B tolerance. Journal of Invertebrate Pathology. 2006;**93**:127-134. DOI: 10.1016/j.jip.2006.05.011

[81]   Crespo R, Juárez MP, Dal Bello GM, Padín S, Calderón Fernández G, Pedrini N. Increased mortality of *Acanthoscelides obtectus* by alkane-grown *Beauveria bassiana*. BioControl. 2002;**47**:685-696. DOI: 10.1023/A:1020545613148

[82]   Hölker U, Lenz J. Solid-state fermentation—Are there any biotechnological advances? Current Opinion in Microbiology. 2005;**8**:301-306. DOI: 10.1016/j.mib.2005.04.006

[83]   Hölker U, Hofer M, Lenz J. Biotechnological advantages of laboratory-scale solid-state fermentation with fungi. Applied Microbiological Biotechnology. 2004;**64**:175-186. DOI: 10.1007/s00253-003-1504-3

[84]   Ibrahim L, Butt TM, Beckett A, Clark SJ. The germination of oil-formulated conidia of the insect pathogen, *Metarhizium anisopliae*. Mycological Research. 1999;**103**:901-907. DOI: 10.1017/S0953756298007849

[85]   Moore D, Bridge PD, Higgins PM, Bateman RP, Prior C. Ultra-violet radiation damage to *Metarhizium flavoviride* conidia and the protection given by vegetable and mineral oils and chemical sunscreens. Annals of Applied Biology. 1993;**122**:605-616. DOI: 10.1111/j.1744-7348.1993.tb04061.x

[86]   Inglis DG, Goettel MS, Johnson DL. Influence of ultraviolet light protectants on persistence of the entomopathogenic fungus, *Beauveria bassiana*. Biological Control. 1995;**5**:581-590. DOI: 10.1006/bcon.1995.1069

[87] Méndez-González F, Loera-Corral O, Saucedo-Castañeda G, Favela-Torres E. Bioreactors for the production of biological control agents produced by solid-state fermentation. In: Pandey A, Larroche C, Soccol CR, editors. Current Developments in Biotechnology and Bioengineering. Current Advances in Solid-State Fermentation. Vol. 7. New Dehli: Elsevier; 2018. pp. 109-121. DOI: 10.1016/B978-0-444-63990-5.00007-4

[88] Muñiz-Paredes FR, Hernández FM, Loera O. Production of conidia by entomopathogenic fungi: From inoculants to final quality tests. World Journal of Microbiology and Biotechnology. 2017;**33**(3):57. DOI: 10.1007/s11274-017-2229-2

[89] Inglis DA, Hagedorn DJ, Rend RE. Use of dry inoculum to evaluate beans for resistance to anthracucse and angular leaf spot. Plant Disease. 1988;**72**:771-774. DOI: 10.1094/PD-72-0771

[90] Berestetskiy AO, Kungurtseva OV. Effects of moisture content in solid substrate on the survival and virulence of *Stagonospora cirsii* mycelium. Mikologiya i Fitopatolopgiya. 2012;**46**(4):280-286 (in Rus.)

[91] Silman RW, Bothast RJ, Schisler DA. Production of *Colletotrichum truncatum* for use as a mycoherbicide: Effects of culture, drying and storage on recovery and efficacy. Biotechnology Advances. 1993;**11**:561-575. DOI: 10.1016/0734-9750(93)90025-I

[92] Kirchmair M, Hoffmann M, Neuhauser S, Huber L. Persistence of GranMet®, a *Metarhizium anisopliae* based product, in grape phylloxera infested vineyards. In: Enkerli J. editor. IOBC Wprs Bulletin. 2006;**30**(7):137-142

[93] Skinner M, Gouli S, Frank CE, Parker BL, Kim JS. Management of *Frankliniella occidentalis* (*Thysanoptera: Thripidae*) with granular formulations of entomopathogenic fungi. Biological Control. 2012;**63**:246-252. DOI: 10.1016/j.biocontrol.2012.08.004

[94] Rosskopf EN, Charudattan R, Kadir JB. Use of plant pathogens in weed control. In: Bellows TS, Fisher TW, editors. Handbook of Biological Control. New York: Academic Press; 1999. pp. 891-918

[95] Barlett MC, Jaronski ST. Mass production of entomogenous fungi for biological control of insects In: Burge MN, editors. Fungi in Biological Control Systems. New York: Manchester University Press; 1988. pp. 61-88

[96] De Vrije T, Antoine N, Buitelaar RM, Bruckner S, Dissevelt M, Durand A, Gerlagh M, Jones EE, Lüth P, Oostra J, Ravensberg WJ, Renaud R, Rinzema A, Weber FJ, Whips JM. The fungal biocontrol agent *Coniothyrium minitans*: Production by solid-state fermentation, application and marketing. Applied Microbiology and Biotechnology. 2001;**56**:58-68. DOI: 10.1007/s002530100678

[97] Jenkins NE, Heviefo G, Langewald J, Cherry AJ, Lomer CJ. Development of mass production technology for aerial conidia for use as mycopesticides. Biocontrol News and Information. 1998;**19**(1):21-31

[98] Hussey NW, Tinsley TW. Impressions of insect pathology in the People's Republic of China. In: Burges HD. Microbial Control of Pests and Plant Diseases. London: Academic Press; 1981. pp. 785-795

[99]   De Cal A, Larena I, Guijarro B, Melgarejo P. Mass production of conidia of *Penicillium frequentans*, a biocontrol agent against brown rot of stone fruits. Biocontrol Science and Technology. 2002;**12**(6):715-725. DOI: 10.1080/0958315021000039897

[100]   Ribeiro-Machado AC, Monteiro AC, Geraldo-Martins BMIE. Production technology for entomopathogenic fungus using a biphasic culture system. Pesquisa Agropecua'ria Brasileira. 2010;**45**(10):1157-1163. DOI: 10.1590/S0100-204X2010001000015

[101]   Adeteunji CO, Oloke JK. Effect of wild and mutant strain of Lasiodiploidia pseudotheobromae mass produced on rice bran as a potential bioherbicide agents for weeds under solid state fermentation. Journal of Applied Biology and Biotechnology. 2013;**1**(2):018-023. DOI: 10.7324/JABB.2013.1204

[102]   Smart MG, Howard KM, Bothast RJ. Effect of carbon dioxide on sporulation of *Alternaria crassa* and *Alternaria cassiae*. Mycopathologia. 1992;**118**:167-171. DOI: 10.1007/BF00437150

[103]   Durand A. Bioreactor designs for solid state fermentation. Biochemical Engineering Journal. 2003;**13**(2-3):113-125. DOI: 10.1016/S1369-703X(02)00124-9

[104]   Mitchell DA, Srinophakun P, Krieger N, von Meienet OF. Group II bioreactors: Forcefully-aerated bioreactors without mixing. In: Mitchell DA, Krieger N, Berovic M, editors. Solid-State Fermentation Bioreactors: Fundamentals of Design and Operation. Berlin Heidelberg: Springer-Verlag. 2006. pp. 115-128. DOI: 10.1007/3-540-31286-2

[105]   Jones EE, Weber FJ, Oostra J, Rinzema A, Mead A, Whipps JM. Conidial quality of the biocontrol *agent Coniothyrium minitans* produced by solid-state cultivation in a packed-bed reactor. Enzyme and Microbial Technology. 2004;**34**(2):169-207. DOI: 10.1016/j.enzmictec.2003.10.002

[106]   Klaic R, Sallet D, Foletto EL, Jacques RJS, Guedes JVC, Kuhn RC, Mazutti MA. Optimization of solid-state fermentation for bioherbicide production by *Phoma* sp. Brazilian Journal of Chemical Engineering. 2017;**34**(2):377-384. DOI: 10.1590/0104-6632.20170342s20150613

[107]   Klaic R, Kuhn RC, Foletto EL, Dal Prá V, Jacques RJS, Guedes JVC, Treichel H, Mossi AJ, Oliveira D, Oliveira JV, Jahn SL, Mazutti MA. An overview regarding bioherbicide and their production methods by fermentation. In: Gupta VJ, Mach RL, Sreenivasaprasad S. Fungal Bio-Molecules: Sources, Applications and Recent Developments. Vol. 1. 1st ed. Oxford: Wiley-Blackwell; 2015. pp. 183-200

[108]   Copping LG, Duke SO. Natural products that have been used commercially as crop protection agents. Pest Management Science. 2007;**63**:524-554. DOI: 10.1002/ps.1378

[109]   Auld BA, Hetherington SD, Smith HE. Advances in bioherbicide formulation. Weed Biology and Management. 2003;**3**:61-67. DOI: 10.1046/j.1445-6664.2003.00086.x

[110]   Alabouvette C, Olivain C, L'Haridon F, Aime S, Steinberg C. Using strains of *Fusarium oxysporum* to control *Fusarium* wilts: Dream or reality? In: Vurro M, Gressel J, editors. Novel Biotechnologies for Biocontrol Agent Enhancement and Management. Netherlands, Dordrecht: Springer; 2007. pp. 157-177. DOI: 10.1007/978-1-4020-5799-1_8

[111] Macko V, Staples RC, Allen PJ, Renwick JAA. Identification of the germination self-inhibitor from wheat stem rust uredospore. Science. 1971;**173**:835-836. DOI: 10.1126/science.173.3999.835

[112] Macko V, Staples RC, Gershon H, Renwick JAA. Self-inhibitor of bean rust uredo-spores: Methyl 3, 4-dimethoxycinnamate. Science. 1970;**170**:539-540. DOI: 10.1126/science.170.3957.539

[113] Lax AR, Templeton GE, Myer WL. Isolation, purification, and biological activity of a self-inhibitor from conidia of *Colletotrichum gloeosporioides*. Phytopathology. 1985;**75**:386-390. DOI: 10.1094/Phyto-75-386

[114] Ley SV, Cleator E, Harter J, Hollowood CJ. Synthesis of (−)-gloeosporone, a fungal autoinhibitor of spore germination using a π-allyltricarbonyliron lactone complex as a templating architecture for 1,7-diol construction. Organic & Biomolecular Chemistry. 2003;**1**:3263-3264. DOI: 10.1039/b308793j

[115] Inoue M, Mori N, Yamanaka H, Tsurushima T, Miyagawa H, Ueno T. Self-germination inhibitors from *Colletotrichum fragariae*. Journal of Chemical Ecology. 1996;**22**(11):2111-2122. DOI: 3316/10.1007/BF02040097

[116] Uspenskaya GD, Reshetnikova IA. Role of pycnidial mucus and some ecological factors in the germination of conidia in the genera Ascochyta Lib and *Phoma* Fr. Mikologiya i Fitopatologiya. 1979;**13**(4):298-301. (In Russ.)

[117] Uspenskaya GD. Ecological adaptation and evolution of fungi. Mikologiya i Fitopatologiya. 1980;**14**(3):259-262. (In Russ.)

[118] Faria M, Martins I, Souza DA, Mascarin GM, Lopes RB. Susceptibility of the biocontrol fungi *Metarhizium anisopliae* and *Trichoderma asperellum* (Ascomycota: Hypocreales). Biological Control. 2017;**107**:87-94. DOI: 10.1016/j.biocontrol.2017.01.015

[119] VanderGheynst J, Scher H, Guo HY, Schultz D. Water-in-oil emulsions that improve the storage and delivery of the biolarvacide *Lagenidium giganteum*. BioControl. 2007;**52**(2):207-229. DOI: 10.1007/s10526-006-9021-9

[120] Paixão FRS, Muniz ER, Barreto LP, Bernardo CC, Mascarin GM, Luz C, Fernandes ÉKK. Increased heat tolerance afforded by oil-based conidial formulations of *Metarhizium anisopliae* and *Metarhizium robertsii*. Biocontrol Science and Technology. 2017;**27**(3):324-337. DOI: 10.1080/09583157.2017.1281380

[121] Shabana Y, Singh D, Ortiz-Ribbing LM, Hallett SG. Production and formulation of high quality conidia of *Microsphaeropsis amaranthi* for the biological control of weedy *Amaranthus species*. Biological Control. 2010;**55**:49-57. DOI: 10.1016/j.biocontrol.2010.06.014

[122] Kolombet LV, Zhigletsova SK, Kosareva NI, Bystrova EV, Derbyshev VV, Krasnova SP, Schisler D. Development of an extended shelf-life, liquid formulation of the bio-fungicide *Trichoderma asperellum*. World Journal of Microbiology and Biotechnology. 2008;**24**(1):123-131. DOI: 10.1007/s11274-007-9449-9

[123] Kolombet LV, Starshova AA, Schisler D. Biological efficiency *Trichoderma asperellum* GJS 03-35 and yeast *Cryptococcus nadoensis* OH 182.9 as biocontrol agents against fusarium head blight of wheat. Mikologiya i Fitopatologiya. 2005;**39**(5):80-88

[124] Sandoval-Coronado CF; Luna-Olvera HA; Arevalo-Nino K, Jackson MA, Poprawski TJ, Galan-Wong LJ. Drying and formulation of blastospores of *Paecilomyces fumoso-roseus* (Hyphomycetes) produced in two different liquid media. World Journal of Microbiology & Biotechnology. 2001;**17**(4):423-428. DOI: 10.1023/A:1016757608789

[125] Pfirter HA, Guntli D, Ruess M, Défago G. Preservation, mass production and storage of *Stagonospora convolvuli*, a bioherbicide candidate for field bindweed (*Convolvulus arvensis*). BioControl. 1999;**44**:437-447

[126] Jackson MA, Payne AR. Liquid culture production of fungal microsclerotia. In: Glare TR, Moran-Diez ME, editors. Microbial-Based Biopesticides: Methods and Protocols. New York: Springer. 2016. p. 1477. DOI 10.1007/978-1-4939-6367-6_7

[127] Mwamburi LA. Isolation and assessment of stability of six formulations of entomo-pathogenic *Beauveria bassiana*. In: Clifton NJ, Moran-Diez ME, editors. Microbial-Based Biopesticides: Methods and Protocols. Vol. 1477. New York: Springer; 2016. pp. 85-91. DOI: 10.1007/978-1-4939-6367-6_8

[128] Quimby Jr PC, Mercadier G, Meikle W, Vega F, Fargues J, Zidack N. Enhancing biologi-cal control through superior formulations: A worthy goal but still a work in progress In: Vurro M, Gressel J, Butt T, Harman G, St. Leger R, Nuss D, Pilgeram A, editors. Enhancing Biocontrol Agents and Handling Risks. Amsterdam: IOS Press; 2001. pp. 86-95

[129] Stephan D, Zimmermann G. Locust control with *Metarhizium flavoviride*: Drying and formulation of submerged spores. In: Koch E, Leinonen P, editors. Formulation of Microbial Inoculants. Belgium: COST; 2001. pp. 27-34

[130] Larena I, De Cal A, Liñán M, Melgarejo P. Drying of *Epicoccum nigrum* conidia for obtaining a shelf-stable biological product against brown rot disease. Journal of Applied Microbiology. 2003;**94**:508-514. DOI: 10.1046/j.1365-2672.2003.01860.x

[131] Jin X, Custis D. Microencapsulating aerial conidia of *Trichoderma harzianum* through spray drying at elevated temperatures. Biological Control. 2011;**56**:202-208. DOI: 10.1016/j.biocontrol.2010.11.008

[132] Norman DJ, Trujillo EE. Development of *Colletotrichum gloeosporioides* f. sp. *clidemiae* and *Septoria passiflorae* into two mycoherbicides with extended viability. Plant Disease. 1995;**79**(10):1029-1032. DOI: 10.1094/PD-79-1029

[133] Jackson MA, Erhan S, Poprawski TJ. Influence of formulation additives on the desic-cation tolerance and storage stability of blastospores of the entomopathogenic fungus *Paecilomyces fumosoroseus* (Deuteromycotina: Hyphomycetes). Biocontrol Science and Technology. 2006;**16**(1):61-75. DOI: 10.1080/09583150500188197

[134] Vorlop KD, Rose T, Patel AV. Encapsulation technology. In: Koch E, Leinonen P, edi-tors. Formulation of Microbial Inoculants. Belgium: COST; 2001. pp. 45-49

[135] Walker HL, WJJr C. Sodium alginate for production and formulation of mycoherbicides. Weed Science. 1983;**31**(3):333-338

[136] Daigle DJ, Cotty PJ. Production of conidia of *Alternaria cassiae* with alginate pellets. Biological Control. 1992;**2**:278-281. DOI: 10.1016/1049-9644(92)90019-A

[137] Gerding-González M, France A, Sepulveda M, Campos J. Use of chitin to improve a *Beauveria bassiana* alginate-pellet formulation. Biocontrol Science and Technology. 2007;**17**(1): 105-110. DOI: 10.1080/09583150600937717

[138] Locatelli GO, Santos GF, Botelho PS, Luna CL, Botelho PS, Finkler CLL, Bueno LA. Development of *Trichoderma* sp. formulations in encapsulated granules (CG) and evaluation of conidia shelf-life. Biological Control. 2018;**117**:21-29. DOI: 10.1016/j.biocontrol. 2017.08.020

[139] Humbert P, Przyklenk M, Vemmer M, Patel AV. Calcium gluconate as cross-linker improves survival and shelf life of encapsulated and dried *Metarhizium brunneum* and *Saccharomyces cerevisiae* for the application as biological control agents. Journal of Microencapsulation. 2017;**34**(1):47-56. DOI: 10.1080/02652048.2017.1282550

[140] Winder RS, Wheeler JJ. Encapsulation of microparticles in teardrop shaped polymer capsules of cellular size. US Patent 6248321 B1. 2001

[141] Winder RS, Wheeler JJ, Conder N, Otvos IS, Nevill R, Duan L. Microencapsulation: A strategy for formulation of inoculum. Biocontrol Science and Technology. 2003;**13**:155-169. DOI: 10.1080/0958315021000073439

[142] Connick WJ, Boyette CD. Granular products containing fungi encapsulated in a wheat gluten matrix for biological control of weeds. US Patent N 5074902. Dec. 24, 1991

[143] Connick WJ, Boyette CD, McAlpine JR. Formulation of mycoherbicides using a pasta-like process. Biological Control. 1991;**1**:281-287. DOI: 10.1016/1049-9644(91)90079-F

[144] Müller-Stöver D, Kroschel J, Sauerborn J. Viability of different propagules of *Fusarium oxysporum* f.sp. *orthoceras* during the formulation into wheat flour–kaolin granules. In: Koch E, Leinonen P, editors. Formulation of Microbial Inoculants. Belgium: COST; 2001. pp. 83-89

[145] Lawrie J, Down VM, Greaves MP. Effects of storage on viability and efficacy of granular formulations of the microbial herbicides *Alternaria alternata* and *Trematophoma lignicola*. Biocontrol Science and Technology. 2001;**11**:283-295. DOI: 10.1080/09583150120035701

[146] Boyette CD, Jackson MA, Bryson CT, Hoagland RE, Connick WJ, Daigle DJ. *Sesbania exaltata* biocontrol with *Colletotrichum truncatum* microsclerotia formulated in 'Pesta' granules. BioControl. 2007;**52**:413-426. DOI: 10.1007/s10526-006-9031-7

[147] Connick WJ, Daigle DJ, Boyette CD, Williams KS, Vinyard BT, Quimby PC Jr. Water activity and other factors that affect the viability of *Colletotrichum truncatum* conidia in wheat flour-kaolin granules ('Pesta'). Biocontrol Science and Technology. 1996;**6**:277-284. DOI: 10.1080/09583159650039467

[148] Shabana YM, Müller-Stöver D, Sauerborn J. Granular Pesta formulation of *Fusarium oxysporum* f. sp. *orthoceras* for biological control of sunflower broomrape: Efficacy and shelf-life. Biological Control. 2003;**26**:189-201. DOI: 10.1016/S1049-9644(02)00130-5

[149] Aybeke M, Şen B, Ökten S. Pesta granule trials with *Aspergillus alliaceus* for the biocontrol of Orobanche spp. Biocontrol Science and Technology. 2015;**25**(7):803-813. DOI: 10.1080/09583157.2015.1018813

[150] Daigle DJ, Connick WJ, Boyette CD, Jackson MA, Dorner JW. Solid-state fermentation plus extrusion to make biopesticide granules. Biotechnology Techniques. 1998;**12**(10):715-719. DOI: 10.1023/A:1008872819909

[151] Quimby Jr PC, Caesar AJ, Birdsall JL, Connick WJJr, Boyette CD, Zidack NK, Grey WE. Granulated formulation and method for stabilizing biocontrol agents. US Patent N 6455036 B1. 2002

[152] Amsellem Z, Zidack NK, Quimby PC Jr, Gressel J. Long-term dry preservation of viable mycelia of two mycoherbicidal organisms. Crop Protection. 1999;**18**:643-649. DOI: 10.1016/S0261-2194(99)00070-8

[153] Quimby PC Jr, Zidack N, Boyette CD, Grey WE. A simple method for stabilizing and granulating fungi. Biocontrol Science and Technology. 1999;**9**(1):5-8. DOI: 10.1080/09583159929857

[154] Zidack NK, Quimby PC. Formulation of bacteria for biological weed control using the Stabileze method. Biocontrol Science and Technology. 2002;**12**(1):67-74. DOI: 10.1080/09583150120093112

[155] Bourdot GW, Hurrell GA, Saville DJ, Leathwick DM. Impacts of applied *Sclerotinia sclerotiorum* on the dynamics of a *Cirsium arvense* population. Weed Research. 2006;**46**(10):61-72. DOI: 10.1111/j.1365-3180.2006.00481.x

[156] De Jong MD, Bourdot GW, Powell J, Goudriaan J. A model of the escape of *Sclerotinia sclerotiorum* ascospores from pasture. Ecological Modelling. 2002;**150**:83-105. DOI: 10.1016/S0304-3800(01)00462-8

[157] Li P, Ash GJ, Ahn B, Watson AK. Development of strain specific molecular markers for the *Sclerotinia minor* bioherbicide strain IMI 344141. Biocontrol Science and Technology. 2010;**20**(9):939-959. DOI: 10.1080/09583157.2010.491895

[158] Amselem J, Cuomo CA, van Kan JA, Viaud M, Benito EP, Couloux A, Coutinho PM, de Vries RP, Dyer PS, Fillinger S, et al. Genomic analysis of the Necrotrophic Fungal Pathogens *Sclerotinia sclerotiorum* and *Botrytis cinerea*. PLoS Genetics. 2011;**7**:e1002230. DOI: 10.1371/journal.pgen.1002230

[159] Wang J, Wang X, Yuan B, Qiang S. Differential gene expression for *Curvularia eragrostidis* pathogenic incidence in crabgrass (*Digitaria sanguinalis*) revealed by cDNA-AFLP analysis. PLoS One. 2013;**8**(10):e75430. DOI: 10.1371/journal.pone.0075430

[160] Zimmermann J, de Klerk M, Musyoki MK, Viljoen A, Watson AK, Beed F, Gorfer M, Cadisch G, Rasch F. An explicit AFLP-based marker for monitoring *Fusarium oxysporum* f.sp. *strigae* in tropical soils. Biological Control. 2015;**89**:42-52. DOI: 10.1016/j.biocontrol.2015.02.008

www.ingramcontent.com/pod-product-compliance
Lightning Source LLC
Chambersburg PA
CBHW081233190326
41458CB00016B/5763